SURVIVE Information Overload™

The 7 Best Ways to Manage Your Workload by Seeing the Big Picture

SURVIVE
Information
Overload™

The 7 Best Ways to Manage Your Workload by Seeing the Big Picture

Kathryn Alesandrini

BUSINESS ONE IRWIN
Homewood, Illinois 60430

Richard D. Irwin, Inc. recognizes that
certain terms in this book are trademarks, and
we have made every effort to print these
throughout the text with the capitalization and
punctuation used by the holder of the trademark.

This publication is designed to provide accurate and
authoritative information in regard to the subject matter
covered. It is sold with the understanding that neither the
author nor the publisher is engaged in rendering legal, accounting,
or other professional service. If legal advice or other expert
assistance is required, the services of a competent
professional person should be sought.

*From a Declaration of Principles jointly adopted by a Committee
of the American Bar Association and a Committee of Publishers.*

Sponsoring editor: Cynthia A. Zigmund
Project editor: Karen Murphy
Production manager: Ann Cassady
Designer: Heidi J. Baughman
Art coordinator: Mark Malloy
Compositor: TCSystems, Inc.
Typeface: 11/13 Palatino
Printer: R. R. Donnelley & Sons Company

Library of Congress Cataloging-in-Publication Data

Alesandrini, Kathryn.
    Survive information overload : the 7 best ways to manage your
workload by seeing the big picture / Kathryn Alesandrini.
        p.   cm.
    Includes index.
    ISBN 1-55623-721-9
    1. Information resources management. 2. Time management.
I. Title.
T58.64.A43 1992
650.1—dc20   91-46561

Printed in the United States of America
1 2 3 4 5 6 7 8 9 0 D O C 9 8 7 6 5 4 3 2

Dedicated to the memory of my father, Carlo Alesandrini; my loving mother, Theresa Alesandrini; and the finest big-picture thinker I know, my husband, Melvin Eddy.

Executive Summary

Thesis: The best way to survive information overload is to see the big picture, because traditional time management practices are too focused and linear to work in the age of information.

The Problem

The explosive growth of information coupled with outdated time management practices has caused information overload in many managers and other white-collar workers. Productivity has suffered—manufacturing productivity grew nearly 1,500 percent faster than white-collar productivity in the past decade—4.1 percent a year versus 0.28 percent, respectively. The majority of Americans feel under pressure to get everything done that they need to do.

The Solution

The solution is to manage your workload by seeing the big picture. The seven ways to do that are summed up by the acronym SURVIVE: Synthesize details, Underscore priorities, Reduce paperwork, View the big picture, Illuminate the issues, Visualize new concepts, and Extract the essence.

- Your ability to see the big picture is more than 60,000 times faster than your ability to process information in traditional ways.
- Plugging into the collective wisdom lets you see the big picture quickly and easily.
- Using office tools, technology, and services to see the big picture helps you reduce reading, make meetings more effective, capture ideas, and organize thoughts.

Top Seven Recommendations

1. The most important thing you can do is to **synthesize details** by adopting a Master Control System or converting your current time management organizer to an MCS. That means you should funnel all important data through the MCS and relate details to major categories.

2. **Underscore your priorities** to make sure you do what's really important. Rather than make lists, you're better off if you map your goals and priorities by category.

3. **Reduce excess** paperwork by eliminating it whenever possible and converting to electronic paper.

4. The way to survive is to **view the big picture** by teaming with others or using technology. Consider teaming with seemingly unlikely candidates such as your customers, or even your competitors. Gear up in using online database services and resources.

5. If you **illuminate the issues,** meetings offer the biggest hope to survive information overload. Convert meetings to active work sessions through proper use of the MCS, presentation graphics, and meeting technology.

6. The way to keep up with learning demands in the information age is to **visualize new concepts** and participate in collaborative learning groups.

7. **Extract the essence** of information with top-down techniques for reading and writing. Cut reading time by subscribing to a clipping service or an online database search service. Tap the collective wisdom through calls to experts or a person in your network who would know the information you need.

Summary

Survival in the information age means looking outside of your own industry so that you can work more successfully within it and plugging into the collective wisdom.

Acknowledgments

I would like to thank the many people who gave of their time to contribute to the collective wisdom contained in this book. In conducting the research for *SURVIVE Information Overload*, I drew from the popular press, academic research, case studies, personal interviews with executives and managers, and input from many others. I extend my sincere appreciation:

—To my fellow speakers in the National Speakers Association who have shown me the way to bring my message to others.

—To my academic colleagues and students who have contributed to my personal growth over the years.

—To my corporate clients and contacts who contributed their wisdom to this book.

—To the thousands of learners in my seminars and speeches for their willingness to learn and, in turn, share their knowledge and experiences with me.

—To my editor, Cynthia Zigmund, for her ability to listen, bring out the best in me, and, most of all, believe in this book.

—To my new friends at Business One Irwin for their dedication and vision, especially publisher Beth Battram, for her willingness to share ideas and brainstorm innovative possibilities.

—To my family and friends who have supported me over the years, especially my sister Janelle Alesandrini for her encouragement and insights early in the project, and my mother, Theresa Alesandrini, for her loving support for this project and for everything I do.

—Most of all, to the person whose love and laughter gave me the inspiration to bring this book to fruition, my husband, Melvin Eddy.

K. A.

Contents

1. Overloaded? .... Synthesize details   1
   with a Master Control System
   Action Plan   18   Self-Test   28

2. Too Much to Do? .... Underscore your priorities   32
   with Vision, Context
   Analysis, and Mapping
   Action Plan   46   Self-Test   56

3. Too Much Paperwork? ... Reduce the excess   58
   with Electronic Paper,
   Color Coding, and
   Visual Organization
   Action Plan   79   Self-Test   88

4. Too Much to Know? ..... View the big picture   91
   with Macros, Online Aids,
   and Broadband Thinking
   Action Plan   111   Self-Test   125

5. Too Many Meetings? .... Illuminate the issues   128
   with Graphics and
   Collaborative Work
   Action Plan   150   Self-Test   155

6. Too Much to Learn? ..... Visualize new concepts   158
   with Chunking and Visual Analogies
   Action Plan   178   Self-Test   185

7. Too Much to Read? ...... Extract the essence   186
   with Top-Down Techniques
   Action Plans   197, 207   Self-Test   211

A Personal Postscript   214

Resource Directory   235

Index   255

SURVIVE Information Overload™

The 7 Best Ways to Manage Your Workload by Seeing the Big Picture

Chapter One

Overloaded? . . .
Synthesize Details
(with a Master Control System)

A forest ranger spots the smallest fire by watching for telltale signs from a lookout high above the treetops, not by running from tree to tree in search of flames.

If you follow the so-called wisdom of traditional time management maxims, you lower your odds of surviving the information age. The problem is that most time management practices are too verbal, too linear, and too focused on day-to-day specifics. To borrow an analogy from George Gilder, time management truisms are like pipes developed for a stream, but an ocean of information needs to get through. It just won't work.

Time management practices are linear and serial, but information comes at you from parallel sources simultaneously—every day the phone rings while a dozen or more pages pour out of the fax; hundreds of papers pile up on your desk every week, and computers spew out another 750 million pages each day in the United States. The extent of the onslaught is mind-boggling:

- Roughly 1.6 trillion pieces of paper circulate through U.S. offices annually. Office paper is growing twice as fast as the gross national product.
- The average worker has 36 hours' worth of work stacked up on his or her desk and only 90 minutes to handle it.
- 15 billion faxes will be sent this year from U.S. fax machines. The total soars to 60 billion if you count faxes issued by computers.

- 12.4 million people are sending 1.2 billion messages via electronic mail each year, up 3000 percent in 10 years.

In the information age, traditional time management practices are not only inadequate, but also downright hazardous to your business survival. Here is the truth about time management myths and what it takes to survive in the information era.

Time Management Myths

More than 50 percent of the current projects in major U.S. corporations are behind schedule, according to a survey of 220 senior executives by Dartmouth Research. That percentage says something about how well (or how poorly) traditional time management approaches are working.

Myth #1. "Focus on your goals and only the information relevant to them, then ignore the rest."

SURVIVE. Information fuels knowledge, which is the lifeblood of the white-collar worker. That knowledge, however, must relate to the entire system; otherwise, the left hand doesn't know what the right hand is doing. Miss one key piece of information about your supplier, customer, competitor, or emerging trend, and you risk not only productivity and success, but ultimately your business survival. An accurate and useful view is, necessarily, a broad view.

Still, how can anyone pay attention to everything that's potentially relevant when there are 50,000 books and 11,000 periodicals published every year in the United States alone. The problem is too much *relevant* information for one person to keep track of unless he or she follows the seven guidelines presented in this book for seeing the big picture.

Successful managers broaden their view to include ideas from outside their normal focus using among other things:

- *A clipping service.* A colleague of mine who runs a multistate training company told me, "In my business, I have to stay abreast of everything that's happening in my field. The clipping service keeps me current in my specialized topic." The typical service employs hundreds of readers who clip items from thousands of publications for a surprisingly affordable fee. There are also "clipping" services for information on television and radio.

Me? I made the mistake of believing time management myths.
Source: © 1991 by Bradford Veley.

- *Summaries, digests, briefings.* A variety of services ferret out the facts so you don't have to. Keep up with business news by listening to a biweekly audiotape news service that covers 150 top business publications. It's especially useful if you commute. Extract the insights of popular business books by simply reading eight-page summaries. Get a handle on a company's background and fundamentals from a digest of one-page overviews. Stay alert to important changes that could impact your business long before the media notices any related trend or opportunity by subscribing to a future-focus briefing.
- *Tailored, online searching.* My favorite way to stay abreast is to search for information online. Computers can search tirelessly,

day or night, through electronic libraries for the information you need. Then they deliver it to your electronic in-basket. As a colleague of mine described it, he has an "automated, electronic information navigator that's out there searching the electronic highways on a global basis" trying to find what's new in his field. When the "navigator" finds it, it lets him know and retrieves the information for him while he sleeps.

• *3-D Information Visualizer.* What better way to keep a big-picture perspective than to add a new dimension to your view. Researchers at Xerox Palo Alto Research Center are doing just that with their Information Visualizer project. Rather than display data in 2-D tables, the project represents data in rotatable, three-dimensional hierarchies.

This rotatable Cam Tree is a hierarchy laid out in 3-D. The root is on top, and the hierarchy proceeds from top to bottom.

Source: Illustration used by permission of the Xerox Corporation.

Myth # 2. "Don't waste time with 'unimportant' people."

SURVIVE. Based on a study of corporate leaders reported in *Harvard Business Review*, top managers ignore this maxim and "see people who often appear to be unimportant outsiders."

These leaders discuss "virtually anything and everything even remotely associated with their businesses and organizations."

The fact is that the seemingly "unimportant" person—whether an employee, customer, supplier, or even competitor—may prove pivotal to a firm's future success. Ryal Poppa, CEO of Storage Technology, holds office hours by setting aside a couple of two-hour periods each week to "free-think." Employees can drop in for quick conversations about unscheduled business matters. When I asked him why he does this, he explained, "I have a philosophy that everybody has good ideas." They have "new eyes" and, therefore, come up with good ideas that might otherwise have been missed. At a recent dinner, one of the company engineers sitting across the table overheard Poppa's informal conversation with others and threw in a suggestion. "An idea came across that was probably worth five million dollars." Not surprisingly, he told me, "It's always worth taking the time."

Beth Lewis Battram took the time to meet individually with every employee when she assumed the helm as publisher of a book company. "You learn things in funny places," she told me. For example, she learned from a part-time secretary the importance to the group of an annual company social gathering.

The tendency to avoid so-called unimportant people can be particularly detrimental to a company's customer service effort. Yet who has time to respond to every question or issue raised by a customer in the age of information? Here are two ways innovative companies meet the challenge:

• *Voice-fax response.* Customers can get answers 24 hours a day from an automatic voice-fax system as I discovered when I called Borland, a software firm, to ask about an upgrade. I was able to access the needed information by phone, select from a menu of choices on the voice system, and receive answers by fax sent automatically by the company's computer within a matter of minutes. A customer support manager later told me that the system is "cheaper than having customers talk with an engineer," although that is not why the company installed it. "The main purpose is to provide an enhanced level of service to the customer." I certainly felt well served.

• *An electronic forum.* Any firm can establish an online bulletin board to communicate with customers, although this has been popular primarily with high-tech firms whose customers know how to access an electronic forum. "No company can duplicate internally the collective wisdom of their entire customer base," a software support manager told me. When a customer asks a question, it's there for everyone in the forum to read, including other customers who share their experiences and offer solutions. The forum is "absolutely the best way to give support." Its one-to-many communication conserves labor yet leaves no one out of the loop—everyone is treated as important.

Myth # 3. "Place a 'Do Not Disturb' sign on your door to limit interruptions so you can be productive."

SURVIVE. "In the ultimate time management system," quips Jack Gordon, "no one talks to anybody, because it takes too much time." In reality, the key to productivity is *collaborative work*, not isolation. Once you begin to see the big picture, you'll find collaborators among your suppliers, colleagues in other areas, customers, and even competitors. Apple Computer, for example, has teamed with Sony and even archrival IBM. Today, however, a "lack of collaboration" is cited as one of the top five obstacles to effective business operations, according to a survey of senior executives in major U.S. firms by Dartmouth Research.

To survive information overload, people can work together by means of:

• *Face-to-face communication.* Collaboration can be as simple as chatting with others. A study by Xerox found that their repairmen learn most about fixing copiers not from company manuals but from "hanging around swapping stories." Instead of breaking up the crowd at the water cooler, organizations should encourage storytelling at informal gatherings.

• *A common metric.* Hewlett-Packard speeds product development by focusing team members' attention on the big picture with a "Return Map," a graph of the time and money it will take to show a return on investment. The map provides a metric that visually integrates marketing, R&D, and manufacturing efforts. If the R&D group wants to make a change, the map shows them what effect it will have on the time it takes to reach profitability. The map helps team members work in true collaboration to reach their common goal.

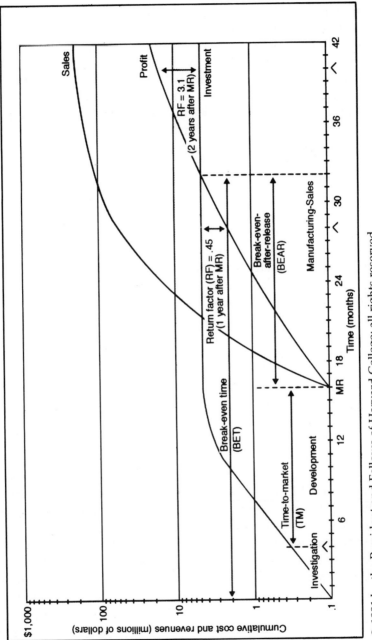

• *A big-picture map.* "Mapping forces you to look at the whole system and recognize that even if your subsystem is optimized, it may not be an optimization of the total system," Motorola manager Brenda Sumberg told me. Mapping was applied to a scheduling process that involved several functional units at Motorola. The effort revealed discontinuities and duplication of effort in the scheduling process. The task was simplified nearly 88 percent from 32 activities down to four basic steps. "If they had automated the process before mapping, they would have automated a dinosaur."

• *Electronic meetings.* You can now link up with others electronically to create or revise documents interactively. "People come out of an [electronic meeting] session saying, 'We've just done 10 hours' work in 90 minutes,' and they can't believe it," says a white-collar worker at Xerox. Collaborative computing allows users to instantly share text, data, and pictures resulting in a level of productivity that often exceeds that of face-to-face meetings. At Electronic Data Systems, meetings are held in a room called the Capture Lab, which provides a personal computer for each participant and a shared public display. The setup "enables groups to actually *do* work in meetings, rather than just talk about it." One company using this technology reports a savings of 71 percent in labor hours and 91 percent in flowtime compared with traditional meeting methods.

Myth # 4. "Carefully plan ahead and schedule your time on a daily basis."

SURVIVE. Successful managers do just the reverse, according to the Harvard study of general managers. They leave much of their day unplanned and spend most of their time in short, disjointed conversations. The truth is that as long as the big picture directs your course, chance encounters that result in short and disjointed conversations can be extremely efficient. Top managers often resolve issues in two-minute "hallway meetings" that save at least 15 to 30 minutes, or 750 to 1,500 percent more time than a hallway meeting takes.

"Your office is your mind. It's not a physical location," says Louis Smith, president of Allied-Signal Aerospace, Kansas City

Division. When I asked him during an interview what he meant, Smith explained that technology allows you to stay in touch whether you're at home, in the car, or on an airplane. He keeps a list of the day's meetings in his pocket but does not overschedule his time. He remains flexible by allowing several hours in the morning and several more in the afternoon to take advantage of opportunities or problems as they arise.

Rather than overplan their day, many managers are relying on:

• *A Master Control System.* A time planner or organizer used in a nontraditional way could function as the MCS, or the MCS could be a high-technology setup. As a centralized source, the MCS synthesizes details in a way that allows you to monitor the big picture, just as a pilot monitors a control panel in the cockpit to fly an airplane. Used in the traditional way, time management tools often make it difficult to see the big picture, because information is scattered across hundreds of daily calendar pages.

• *A mental outline.* The key to seeing the big picture is to group information into a hierarchy of related categories, then monitor the categories with your Master Control System. "The best way to simplify complexity is through classification," says Dan Burrus, a leading science and technology forecaster. He told me that he spent a year developing a taxonomy of high technology early in his career, because "I was fogged in with information. It was too complex." Now he updates the categories as the big picture changes. Computer outliners make it easy to view or change categories as needed. As a "desk accessory" on your computer, an outliner can be called up with one key press to check or modify your outline while you're working on other things.

Myth # 5. "Never handle a piece of paper more than once."

SURVIVE. "I haven't figured out how to do that," a publishing executive admitted to me. "Every day we have two or three mail deliveries. I'm barely able to skim everything by evening much less act on each and every item. And the problem is compounded by those of us who travel a good bit."

Here's why this time management dictum sounds valid on the surface but doesn't work in the age of information. If you put off handling an item until you have time to respond or take care of it, that item adds to the growing number of details you cannot monitor from your big-picture perspective. Before you know it, the pile of details grows to such proportions that you feel overloaded. Consider:

• *Immediate integration.* Spend just a few minutes each day integrating paperwork into the organizational structure. When I don't have time to respond to a particular item, I file it with related items and make a notation in my MCS to take care of it later. That's where a Master Control System is essential. Otherwise, notations to "do it later" end up ignored or lost in the clutter.

• *Files that fit the big picture.* Managers delegate the task of filing to the staff, but the best person to make sure that files fit the big picture is the manager. Since it's important to funnel information through major categories, file categories should match the big picture. With information organized appropriately, both managers and their employees escape the trap of time-based, linear thinking and enter the realm of big-picture thinking.

• *Electronic paper.* Companies are using electronic imaging technology to convert paper documents into electronic paper, which employees can "handle" as much as they want without wasting time. Electronic paper has "totally changed the way we do business," says Robert McDermott, CEO of the United Services Automobile Association (USAA). Each day some 30,000 pieces of mail never leave the mailroom. Instead, an exact image of the correspondence is placed electronically in the customer's file, giving everyone in the company instant access to the information.

Seeing Your Way Out of Information Overload

Seeing the big picture is your key to survival in the information age. If you're suffering from information overload (and who isn't?), you can see your way out—more quickly than you might think. The secret is to manage details by seeing the big picture.

The trouble is, it's not easy to keep an eye on the big picture while the information explosion hits crisis proportions around you. If you're overwhelmed already, it seems like a luxury—if not an impossibility—to spend time pondering the big picture. But that's exactly how top leaders, CEOs, and other successful managers triumph over too much paperwork, too many meetings, and too much to do. They literally *see* their way out of information overload.

It's easy, however, to get bogged down with the details. Face to face with a burning tree, we're blinded by the flames. That's when we make mistakes, miss opportunities, and fall victim to information overload. Information becomes the enemy rather than an ally.

Those who lose sight of the big picture will end up getting burned, like John.

Is This You?

John doesn't have time to plow through an important, but thick, report on top of the growing pile of memos, articles, forms, correspondence, and other reading material on his desk. He does the next best thing—he flips through the report while talking on the phone so he can honestly say he's "looked it over."

John's day is fragmented into 11-minute segments (the national average) for dealing with any one issue. He gets through only five or six pages during a "free" segment since he reads at the average adult rate of 250 words per minute. Most of the 145 pages that come in during the week wind up in the mass of clutter on his desk.

Meetings and phone calls account for much of the day. Like most managers, John spends nearly 50 percent of his time in meetings, but he doesn't realize meetings take that much of his time. Only 8 percent of his time goes to analysis and decision making. His mistaken impression is that the figure is much higher. He rushes in late to the afternoon staff meeting because he had to search for the agenda buried on his desk top.

The meeting drags on yet is unproductive. A staff member missed an important deadline and acts unaware of the slowdown it will cause for others. John explains how the task fits into the bigger picture. Others echo their surprise that projects relate to some invisible master plan.

Late in the day, John spends some time preparing for a trip abroad to help with the firm's international expansion. He's supposed to learn about the country's culture and have a rudimentary command of the language. But he can barely keep up with day-to-day demands, let alone find time to immerse himself in a major learning project.

By the end of the workday, John pauses to consider the situation. He barely made a dent in the growing piles. He has literally thousands of pages waiting for him to read and digest. His staff has fallen behind despite a flurry of overtime work. Tomorrow will bring more paperwork, more missed deadlines, and more pressure. And he wonders what's wrong as he enters the *danger zone* of information overload.

Telltale Signs of Information Overload

You most likely suffer from information overload if your work requires you to read reports, keep up with trade publications, write proposals, attend frequent meetings, travel, learn new skills, or monitor financial data and current events. You are not alone. Almost every white-collar worker suffers from information overload to some degree. These are some common symptoms of overload:

- You spend too much time dealing with trivia.
- You ignore some vital information completely.
- Sometimes you fail to recognize important information until it's too late.
- You often respond to information in habitual or typical ways (like throwing it away) without considering the consequences.
- You "pigeonhole" information according to arbitrary groups.

The problem occurs since information isn't knowledge. Yet knowledge is what we seek. Richard Wurman dubbed the problem and his book *Information Anxiety* when "information doesn't tell us what we want or need to know." Naisbitt, writing in *Megatrends* and *Megatrends 2000*, defined information overload as our inability to convert information into knowledge. "We are drowning in information but starved for knowledge," he wrote. "Uncontrolled and unorganized information is no longer a resource in an information society. Instead, it becomes the enemy of the information worker."

The irony of information overload lies in the vicious cycle it engenders. What does the typical overloaded worker do? Too often we either ignore information altogether or frantically pore over more to find needed answers. We call another meeting, make more phone calls, read yet one more report.

It's Like "Drinking from a Fire Hose"

In my study of the problem, I've found nearly everyone suffers from information overload to one degree or another. Here's just a sampling:

President of an R & D corporation

"The volume of information is staggering in the analytical business. I often feel as if I'm drinking from a fire hose. I try to be selective, to delegate research and read only articles by journalists and academicians I know are credible. I won't spend more than 30 minutes with a news magazine, and limit my newspaper reading to news, business, and sports. But from 5 P.M. to 7 P.M., after everybody else has gone home, I'm usually still here wading through stacks of paper. When I'm finished at the end of the day, I know there are dozens of interesting things I never got to. It's very frustrating."

President of a real-estate firm

"I have to be aware of everything that could influence the purchase of a house. That's why I subscribe to more than 100 real-estate newsletters, legal journals, and economic reports. I read 28 different Sunday newspapers from across the country. I also review every listing, escrow, and sales report from the 1,400 people in my six offices. I spend two hours every night studying data sent to the fax machine in my house."

Executive of a public utility

"I receive more than 100 documents, letters, journals, and reports each day. I can't disregard anything because, in this job, the political information is just as important as the technical. With millions of bits of information passing by me, I've given up trying to memorize statistics. I just try to understand concepts. I have to deal with 10 times more information than a decade ago. Ten years ago, who thought about the Earth getting warmer? Now there's an entire body of knowledge on the greenhouse effect. I used to garden, but who has time for a hobby anymore?"

Television anchorman

"A decade ago, you had a daily newspaper and two wire services. Today there's much more news. Between reading the three major news magazines, I watch the morning news, spend an hour reading the *Times*, listen to [news radio], monitor CNN throughout the day, and look at all three major nightly network news shows. A producer saves me a lot of time by sorting hundreds of stories into 22 computer baskets. . . . Where did people ever find the time to go rip stories off a wire machine?"

City politician

"Sometimes I sit there, knowing I have an important vote to cast the following day, and say to myself, 'I don't know what this means.' The Community Redevelopment Agency budget was so dense I finally had to send the agency a list of questions. The response totaled 150 pages, twice the size of its budget."

Engineering VP

"When I got into computer analysis of finite elements in (airframe) structures in the mid-50s, there were only three professional publications. Ten years later, there were 10,000. Today, there is so much technical data that some magazines do nothing but synopsize documents in other journals. I have 8,000 engineers monitoring every advance in aircraft engineering and design, but I still must go through about 6 inches of technical mail every day."

A New Look at the Overload Problem

Let's rejoin John, the typical manager, as he sinks into his easy chair for a quiet evening at home. He flips on the TV to watch what amounts to 30 pictures per second (that's the U.S. standard) flash before his eyes. He views nearly one half million images that evening. If the adage is true that each picture is worth a thousand words of information, he relaxes by seeing the equivalent of a half billion words—enough to fill 2,500 volumes averaging 500 pages each!

Numbers and words quickly saturate our perceptual capacity, but we actually enjoy viewing visual information in the form of graphics, photos, images, moving pictures, and other media. If you are like most people, you generally feel *entertained*, not overloaded, by watching television. Yet an evening of television bombards you with nearly a *million* percent more times more information than an evening with the company's financial report.

The Wrong Medium Can Cost You Business

Overloaded workers fail to notice important trends, lose track of the competition, and miss new opportunities. What's worse, most of us don't recognize these symptoms of information overload until it's too late.

When Jim Manzi, president of Lotus Development Corporation, monitored the company's performance, he faced thick monthly reports that came in the form of—not surprisingly—a stack of Lotus 1-2-3 spreadsheets. Like most busy executives, he had no time to ponder every number in every table. Unfortunately, he missed an important detail buried in a table on page 24. It showed sales slipping on one of their products.

The executive's subordinates assumed he was taking care of the situation. They had done their job by sending him all the numbers. However, sales of the product continued to slide unheeded for several months. The problem grew to the point where it captured everyone's attention. By then, however, precious time had been lost in taking corrective action in the early phase when it counts most.

After the incident, Lotus began converting its spreadsheet data into graphics for monthly executive reports. The conversion process turned out to be almost effortless since the company's own graphics programs plot the spreadsheet data automatically. Now the executives stay better informed with less time and effort, because it's easier to get the picture.

"Since we switched from spreadsheet reports to chart reports," explains Manzi, "senior management has a much better view of what's going on at all levels of the company. Key information isn't buried in a single spreadsheet cell on page 24; any trends are instantly visible from the very beginning." The best part is that "people are actually reading the report because it gives them critical information in a clear, digestible, actionable format."

The Right Medium

The right medium helped the Lotus management team *survive information overload* by using visualization to keep an eye on the big picture. Part of the power of visualization lies in its ability to simplify information.

Feeling smug that he "boiled an ocean of paper down to a single sheet," the vice president of finance at one of the 10 largest companies in Mexico found that visualization solved his overload problem. "Of all the frustrations of business life, surely one of the most aggravating and persistent is the flood of paper. Until a year ago, I used to update my mental portrait of my company by wading through a 100-page monthly budget report full of data on the corporation, the divisions, the profit center, and the products.

"To round out the picture, I also slogged through a series of smaller reports on collections, bank loans, and the like. These added perhaps 50 pages to my pile.

"Now I get a better picture from just *one sheet of paper*. It has 20 small graphs on it."

The executive's single sheet of paper condenses a voluminous amount of data into a manageable view of the company that shows how present performance relates to both the past and the future. "The combination of charts gives us a concise overview of our business," he explains. "In many cases it allows us to make decisions without further detail."

Seeing Is Deceiving

Escaping information overload is not simply a matter of visualization, because that could work against you. Here's why: One minor fact about your business or the competition, if portrayed vividly enough, can outweigh a dozen more important facts. Advertisers often show testimonials for this reason. They know that most people are influenced more by seeing one happy customer than by hearing an abstract statistic from the competition about a significant percentage of dissatisfied customers. A problem occurs when minor, but vivid, information deceives you into forming an inaccurate big picture. It happens more often than many people realize.

I conducted a marketing research project some years ago during which several top managers fell victim to the "seeing is deceiving" syndrome. As a consultant, I was brought in to conduct focus groups with prospective customers around the country to test consumer reaction to one of the company's newly redesigned products.

A roomful of managers from the company attended the focus group session held in Washington, D.C. and observed customer reactions from an adjacent room through a two-way mirror. As planned, most of the people in that particular session had never used the product or one like it. Since everything was new to them, they were very impressed and offered little or no criticism.

In contrast, participants reacted quite differently in a later session held on the West Coast that the managers did not attend. Most of these participants were experienced users of the product or one like it and, therefore, had high expectations. They were quick to point out shortcomings and areas for improvement. To compound matters, the product demonstration did not go well. Several capabilities could not be shown because the product was not functioning properly.

Since top managers saw the positive response in D.C. with their own eyes and heard reports of the critical feedback secondhand, I was not surprised when they largely discounted the negative reaction and attributed it to the product breakdown. They decided not to make additional major alterations before releasing the product, which, unfortunately, was not as well received nationwide as it could have been. I learned a lesson: to make sure that managers witness the full range of customer reactions with their own eyes, either directly or on tape, before they draw conclusions from focus group research.

"No One Pays Attention to the Big Picture"

This warning comes from Dr. Rosabeth Moss Kanter, editor of the *Harvard Business Review*. "Each player has a piece of the action, but no one makes sure they are all working together on the same team. The materials manager does not seem to care about complaints from the shop floor. The marketing folks ignore the sales team, and so on."

The net effect is a lack of productivity among white-collar workers. Service companies—where 85 percent of white-collar workers are found—have posted disappointing gains in productivity, a red flag for future economic growth.

- In recent years, productivity in nonmanufacturing, which includes services, mining, and construction, has actually *declined*.

- Yet the service sector accounts for 60 percent of the gross domestic product worldwide and most of the new jobs created in the industrial countries. In the United States alone, some 8 of 10 workers are employed in the service sector.

- In the last decade, manufacturing productivity grew nearly 1,500 percent faster than white-collar productivity—4.1 percent a year versus 0.28 percent, respectively.

To succeed today, in fact to *survive*, you need to boost your productivity through big-picture thinking—a top–down view of your work. To achieve it, you need a control system that consolidates your diversity of responsibilities into a manageable "handle." The tool that allows you to streamline your view by synthesizing the many details into one manageable focus is the Master Control System.

ACTION PLAN: Devise a Master Control System

Step 1—Set Up a Master Control System (MCS) or Convert Your Current Planning Device to an MCS.

Just as a pilot uses a master control panel to monitor and fly an airplane consisting of many complex subsystems, you can keep track of yourself, your staff, and your company with a Master

Control System (MCS). Here's the reason: It funnels an unmanageable collection of dates, times, numbers, names, and other needed data into one centralized resource available at your fingertips. A Southern California stockbroker traded the more than 300 slips of paper floating around his desk for an MCS and increased his productivity an impressive 12 percent, which translated to thousands of dollars in additional income. According to figures provided by Time/Design, the MCS can save you an hour and a half each day.

A Master Control System may consist of anything from an individual time planner if properly used to a companywide computerized Executive Information System. As an individual, the following three-point plan will help you devise or adapt your own MCS in just a few hours.

Categorize your 12 most important activities. Try to boil down your areas of responsibility and concern—personal as well as professional—to roughly a dozen general categories. This step requires you to take a big-picture view of your life. Do your activities seem too complex to classify in only 10 or 12 ways? "The best way to simplify complexity is through classification," recommends Daniel Burrus, a leading science and technology forecaster who created a taxonomy to simplify the complex world of high technology.

Categories should include major projects, activities that you monitor, people with whom you deal, and information you need. Any area that consumes hours of your time on a weekly basis becomes a candidate for a separate category.

When I first categorized my activities, they included Consulting, Teaching, R&D, Writing, Public Speaking, Employees, Property, Financial, Personal, Organizations, and two additional categories for major consulting projects. I later split Organizations into three sections, because I spent so much time on the board of directors of one group and as an officer in another. I also divided the Personal category into several separate designations and ultimately merged others into a new University category.

A marketing director who handles several major product lines told me his 10 main categories consist of Collateral for one major product line (brochures and product information), Promotions,

Collateral for the other line, Training, Software, Internal Communications, Public Relations, Finances (to track the financial picture of the company), Personal (his own finances, exercise schedule, etc.), and Ideas.

Chances are you will revise your categorization scheme as you begin to use it. However, by starting with a dozen or so broad categories, you'll be able to relate specifics to one of these categories.

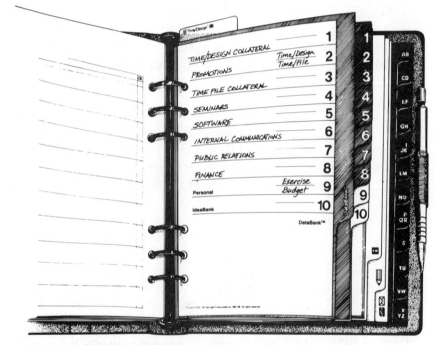

Activities are divided into 10 major sections.
Source: Courtesy of Time/Design, Inc.

Use a hand-held time planner that provides daily, weekly, monthly, and yearly views. According to an old Chinese proverb, "It is not enough to come to the river intending to fish. You also have to bring a net." For most people, it makes sense to invest in a traditional time planner or enhanced calendar system that can be carried around with you in a hand-held binder. Popular systems are available from Day-Timers, Day Runner, Filofax, Franklin, Time/Design, and others. See the resource directory for de-

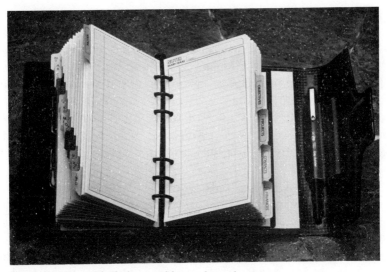

Time planner with daily, weekly, and yearly views.
Source: Courtesy of Day Runner, Inc.

tails. Each system has its own particular approach, but they all help you keep track of goals and activities on a daily, weekly, monthly, and yearly basis.

The planner should feature:

- A **loose-leaf ring** arrangement to allow you to customize and supplement with new pages.
- A **calendar section** that provides weekly, monthly, and yearly views. These views help you see the big picture.
- A numbered (or lettered) **project data section** with a dozen or so tabs. This is the heart of the MCS. You customize it by adding pages for project planning, meeting notes, checklists, other project information, and relevant contact data and phone listings. A paper key at the beginning of this section provides an index to relate each tab to a specific project or category.
- A **general section** with compartments for photos, computer disks, credit cards, ID cards, stamps, business cards, etc.

Each category that you identified as important becomes a separate tab in the project section of your planner. For example, if you are working on three major projects, start separate sections for each. Similar information can be grouped into a single section. Say, for example, you have six people reporting directly to you. Start a Staff section with six separate pages to keep track of each person's responsibilities, tasks, deadlines, and so on. Commitments you've made to your boss can be kept in a separate section, labeled Boss or Supervisor.

You can start using a system in as little as an hour or so. However, the more time you invest in mastering the MCS, the bigger the payoff in productivity and time savings.

Converting your current system. If you already use a time planner, convert it to a Master Control System. The main problem with most planners and calendar systems is that they focus your attention on daily tasks and details. Important information is difficult to pull together into a big picture, because it is spread across dozens, if not hundreds, of daily pages. The first thing to do to convert your current system is to focus on categories rather than the daily calendar. Make only cryptic notes on the daily calendar that refer you to major categories. The guidelines in Step 2 will help you use your planner to see the big picture.

Incorporating technology. If the MCS seems like a "low-tech" solution in today's high-tech world, you can either incorporate high technology into a paper-based planner or adopt an exclusively high-tech system.

Technology expert Dan Burrus relies exclusively on electronic tools. He carries a notebook computer the size of a standard piece of paper (8 1/2 × 11) that's only an inch thick and weighs a few pounds. When I asked him why he didn't use a standard binder, he explained he has eliminated paper partly due to its bulk and weight but primarily because it is static. "Static information is basically dumb and stupid. You have to find it, it can't find you. Dynamic information, though, you can tell to find you when you need it. You don't have to go looking for it." Once he enters notes in his notebook computer, he enters a key phrase or topic for the information. Then whenever he needs information on that topic, the notes will be retrieved instantly.

The electronic MCS is not for everyone. One drawback is the limited life of the battery. Another is that it is not always appropriate to use a computer. "It's not polite to whip out a computer in a meeting and take notes by typing," one marketing exec told me. "You can't pay attention to the speaker if you're busy concentrating on typing"—or, at least, that can be the impression.

Before adopting a binder-based MCS, I used a laptop computer with time management and project management software. In theory, the exclusively high-tech setup should have boosted my productivity, but in practice it didn't work that way for me. The main problems were lack of instant access and portability. I want to be able to flip to a page in my planner as I'm heading down the hall to catch a plane. I got eyestrain looking at the little screen. I'm awaiting a "palmtop" computer with a brighter screen, longer battery life, and quieter keys than today's models. For now, I use a combination of high and low tech. Here are a few tips for combining the two:

- Photo-reduce computer printouts of relevant data such as phone numbers, hole-punch, and add as pages in the appropriate section of the planner.
- Buy prepunched packets of computer paper from the manufacturer of your time planner and use it to print out charts and diagrams, lists, objectives, addresses, and other summary data.
- Add a small calculator that can double as an electronic address book. Many planners provide a special place for a small calculator. If not, get a miniature calculator and slip it into the plastic sheet provided for credit cards.
- Check with the manufacturer of your planner to see if it offers related software products such as contact management software and calendars, since several now do.
- Store computer disks containing critical information in plastic pages designed for disk storage.
- Add an outliner to your personal computer and duplicate your MCS categories in it. By handling the outliner as a desk accessory, you can call it up instantly to access your hierarchical list of categories while working on other software.
- Begin using a "personal information manager" (PIM) software product if you don't already. In a fax-poll conducted by *PC-Computing,* this type of software was

among the top five new products of interest to their readers. "Readers may not be sure what fits in this category—they named everything from *Windows* utilities to the personal-finance manager *Quicken* [neither of which is a PIM]—but they're ready to try it." An example of PIM software is *Agenda* from Lotus. It lets you view your activities, projects, contacts, and other information by categories. Then add relevant printouts to your MCS.

Funnel *all* important data through the MCS. Summarize important information in the MCS as a central resource and store related materials in corresponding office files. From this point forward, channel **all pertinent information** through your Master Control System by referencing key people, commitments, plans, and resources.

The MCS reduces overload, because you can find reference to everything in one place. Just as a pilot need only glance at the cockpit gauges rather than run around the plane to check the oil, fuel, and other critical data, you can consult your MCS rather than waste time figuring out where you'll find important information. The MCS either contains it or quickly guides you to the appropriate source.

It's important to funnel information through the categories you have set up. By relating information to main categories, you escape the trap of time-based linear thinking and enter the realm of hierarchical, big-picture thinking.

Step 2—Follow These Six Tips to Maximize the MCS

Whether you've used a planner for years or started today, the following guidelines help you use this important device to manage details in light of the big picture.

1. Relate details to major categories. To make the most of your MCS, relate details to some larger project, area, or category. Does your desk or briefcase collect little slips of paper with phone numbers, messages, and appointments? Clear the clutter by entering significant items in your planner.

- Keep a sheet of paper at the end of each project section for project-related phone numbers. Then if you need to call Ann about project number 2 next Tuesday, write "Call Ann—P2" on the calendar page or task list for Tuesday. The reference "P2" directs you to the corresponding project section, which includes Ann's phone number as well as important project information. As you discuss the project with Ann, you can add notes on your conversation directly into the project section where they won't get lost.
- Gather receipts, credit card slips, and the like into a Financial receipt envelope. If related to business travel, the envelope can go in a Travel section along with data about travel agents, trip itineraries, airlines, and frequent flyer numbers.
- Store details that don't fit into any larger category, at least not initially, in a miscellaneous section. Business cards, interesting quotes, and good ideas can be held temporarily in a general section as long as you integrate them as soon as possible. Does a new quote illustrate a point in your upcoming presentation? If so, move it or refer to it in the presentation planning section.
- Use the flexibility of personal-information-manager (PIM) software to assign information to relevant categories that you can later access from a variety of perspectives. Here's an example based on a popular PIM. Type in your task to "review Sarah's plan for the Wells proposal," and the PIM software assigns it to these categories: Tasks, Reading, Sarah, and Wells proposal. Then when you ask for a list of things you need to read, this task will pop up. Or if you know you'll see Sarah at 2:00, you can look quickly at all the information you've put in the Sarah category, so you can be sure you're aware of everything you might want to discuss with her when you meet.

Whether or not you use technology, the point is to see the big picture by relating details to larger categories. Have you ever called someone only to forget the key issue or question you had intended to discuss? It happens when you become so immersed in detail that you lose perspective. The value of the MCS lies in its ability to help you maintain a vision of the big picture while tracking specifics.

2. Use only one calendar. A common error is to have one calendar on your desk at work, a smaller one that you carry around with you, and a third at home for recording personal business. With only one calendar, you force all date-related information through your hand-held binder, allowing it to function as a true Master Control System. (See tip 5 for alternatives to carrying a bulky planner everywhere you go.)

- Destroy all but *one* planning calendar. Decorative calendars are OK as long as you don't write on them.
- Use different colors to record important activities of others: blue for your assistant, green for your spouse, brown for the kids.
- Resist the temptation to keep a separate social calendar. Channel everything through your MCS. Color coding and abbreviations make it possible to collapse several calendars into one.
- Schedule your own calendar as a rule. If you *must* allow your secretary or someone else to schedule time for you, allow only 2–3 hour blocks to be scheduled rather than the entire day. Mark off the blocks on your own calendar that you "give" to someone else to schedule. Then you won't make commitments for those times without checking availability first.

3. Enter information only once. People who repeatedly write the same information in more than one place in the MCS often quit using a planner altogether, because it takes more time than it's worth. They find themselves copying unfinished tasks from one day to the next or moving appointments from monthly to weekly to daily calendars.

One sales rep transferred phone numbers from Monday's unmade calls to Tuesday, Wednesday, and so on to the end of the week. Then she would start the entire process again the following Monday. She solved the problem by listing project-related phone numbers in Prospecting and Customer categories. Now, she marks off time for prospecting on her daily calendar and flips to the project section for names and numbers of people to call.

4. Maintain consistency between the MCS and files. Coordinate categories across your MCS, office files, computer files, and personal files. If the office files do not already have categories that correspond to those in your planner, label them now for later use. Although the office files will have many subcategories, the headings for files which you use should match the headings in your planner. Every project section in the MCS should have a corresponding location in the files where project-related materials can be housed.

If the unthinkable event happens—you lose your MCS—you can reconstruct it from the supporting material you have filed. Remember, the MCS functions like a funnel: it summarizes projects, plans, commitments, and other information described more fully elsewhere. By the way, new MCS users do misplace their planners, especially during the first week or so. However, once it becomes an integral part of your work, the MCS stays with you like a wallet or purse.

5. Carry the MCS with you at all times. Keep your planner—or a space-saver stand-in—with you at all times to maximize productivity and minimize overload. Here's what happens when your MCS remains in a desk drawer. You scrawl phone numbers on scraps of paper, record messages on envelopes, and draw napkin graphics for a great idea that too often ends up in the trash. Important information falls through the cracks.

- Use a space-saver substitute for the MCS during social engagements and other times when you don't want to carry a planner. A small pad of "sticky notes" lets you jot down ideas, numbers, and other information wherever you are. Later, it's easy to post these notes in the related section of the MCS so the pertinent information is quickly integrated into the big picture. Be careful, however, that you do not overuse such notes.
- Check with the manufacturer. Several planners include optional satellite systems to save space, but the danger is that you'll start writing on the satellite calendar and violate tip 2: "Use only one calendar."

- Bring your MCS on vacations to keep track of good ideas that invariably pop up when you're relaxing. An added benefit is that it affords a reassuring look at the big picture so you can relax and enjoy your vacation.

6. Begin each category section with a top–down view. Each category in the MCS should begin with its own "big picture." That way, you'll start to think along the right lines every time you flip to that category. The next chapter covers specific tools and techniques to accomplish this task.

- Take the first step by simply including an index of subcategories at the beginning of each category section.
- Write a simple outline of the section. As I wrote this book, the table of contents and a simple map were on the first page of the book's section in my MCS. Every time I referred to the section during phone conversations or other tasks, I was reminded of how that task related to the book's structure.
- House the corporate mission statement, company calendar, or other key company data in an appropriate section of your MCS. Some planners offer customized forms, or your company can print special forms to fit a specific job, department, or the entire company. Distributing these customized forms to all managers assures that everyone is operating with the same big picture.

The point is to include a top–down view that will elicit big-picture thinking.

SELF-TEST: Are You Headed for the Danger Zone?

To discover the extent to which overload currently limits your productivity and effectiveness, take the following self-test and find out how to use this book for maximum benefit.

	Yes	No
1. Can you find any document on your desk in a matter of seconds?	____	____

2. In the last week, did you spend more
    than 10 minutes trying to find some-
    thing in your office?   _____   _____

3. In the last week, were there any papers
    on your desk, other than reference ma-
    terials, that you did not refer to?   _____   _____

4. In the last month, did you attend a
    meeting that lasted longer than sched-
    uled, yet was unproductive?   _____   _____

5. In the last three months, did you fail to
    answer an important letter because you
    misplaced or ignored it?   _____   _____

6. In the last three months, were you sup-
    posed to attend a seminar or learn
    something new but couldn't find time?   _____   _____

7. Do you bring paperwork home from the
    office more than once a week?   _____   _____

8. Do you feel that you have more work to
    do and more meetings to attend than
    anyone could possibly handle?   _____   _____

9. Can your assistant find any document in
    the office files within five minutes?   _____   _____

10. Do staff members meet project
    deadlines without a flurry of overtime
    work at the last minute?   _____   _____

Scoring

Questions 1,9,10: 1 point for each "no."
Questions 2–8: 1 point for each "yes."

What Your Score Means

Score of 0   Congratulations! You're safe. Skim this book
                  for any additional pointers and ideas for the
                  future.

Score of 1–2   Approaching the Danger Zone. Skim the
                  book, focus on the sections that are most

helpful to you, and save the book for future reference.

Score of 3–5   Entering the Danger Zone. Note your "wrong" answers and check the table of contents for the appropriate chapters to zero in on after reading through the entire book.

Score of 6–10   In the Danger Zone of Overload! Your ability to get the job done is blocked by ineffective strategies for handling projects, people, information, and time. Read and study *SURVIVE Information Overload*—it will change your life.

Escaping Information Overload Step-by-Step

More than ever, white-collar workers are being pressured into overload. The forces are many: increased competition, globalization, new technologies, and corporate restructuring that eliminated layers of managers but not the work they used to do.

In the following chapters, you'll discover how you can respond to these pressures and follow step-by-step plans to map priorities, eliminate visual clutter from the work environment, chart progress, and keep an eye on the big picture. Follow the step-by-step *Action Plans* provided at the end of each chapter to escape the crushing burden of information overload. You'll learn to tap your ability to see the big picture in making information your ally. Finally, you'll discover how CEOs and other top managers conquer information with big-picture thinking.

You can start by taking the Self-Test at the end of each chapter. Then score yourself to see whether you'll benefit from studying the chapter or can skim and skip ahead.

This book will not tell you everything you need to know to be successful in this age of information. It does provide you, however, with valuable tools for handling the information explosion. The specific techniques and practical strategies offered here are easy to learn and master. Best of all, they work. You'll *see.*

Coming up. You have taken the first two steps toward escaping information overload. You devised a Master Control System and learned basic principles of how to use it. Turn to the next chapter to find out how you can go beyond the basics and best use the MCS to "underscore your priorities."

Chapter Two

Too Much To Do, Too Little Time? . . .

*Underscore Your Priorities
(with Vision, Context Analysis,
and Mapping)*

How often I saw where I should be going by setting out for somewhere else.

R. Buckminster Fuller

The real voyage of discovery consists not in exploring new landscapes, but in having new eyes.

Wally "Famous" Amos

Time is a most versatile resource. It flies, marches on, works wonders, and will tell. It also runs out. As one study put it, we're caught "between a clock and a hard place." Eight in 10 Americans, 78 percent, feel that time moves too fast for them. Over half, 54 percent, say they're under pressure to "get everything done that you need to." Despite the fact that a majority, 52 percent, make time management lists or calendars to help organize their time, the average American over a lifetime will waste five years waiting in line, three years attending meetings, one year searching for things in the home or office, and eight months opening junk mail.

Once time was money. Now "time could end up being to the '90s what money was to the '80s," predicts *Time*. In fact, timeliness contributes more to success than money according to a McKinsey study. It reports that, on average, companies lose 33 percent of after-tax profit when they ship products six months late, as compared with losses of 3.5 percent when they overspend 50 percent on product development.

"Time stays long enough for those who use it," Leonardo da Vinci once said. The difficulty in using time well today lies in keeping an eye on the big picture while grappling with day-to-day demands punctuated by fax machines, cellular phones, and satellite pagers.

How Effective Executives Manage Their Time

Eighty-five percent of executives believe working overtime is not cost-effective. The 70-hour workweek is a myth; most managers lose their energy and productivity after about 10 straight hours of work. On the average, Americans who earn over $50,000 a year spend 46 hours a week on the job. Entrepreneurs put in somewhat more, 52.6 hours a week. So how do top managers get everything done with their limited work time?

Surprisingly, effective executives rarely follow traditional advice from time management experts to limit interruptions, avoid "unimportant" people or topics, and carefully plan ahead. Based on a study of corporate leaders reported in *Harvard Business Review*, top managers:

- Leave much of their day unplanned.
- Often see people who appear to be unimportant outsiders.
- Discuss "virtually anything and everything even remotely associated with their businesses and organizations."
- Spend 70 percent or more of their time with others.
- Joke and kid about non-work-related topics. The humor is aimed at others in the organization or industry. Nonwork discussions generally concern family, hobbies, or outside activities.
- Spend most of their time in short, disjointed conversations.

Chance encounters that result in short and disjointed conversations can be both effective and efficient. The following is a typical example of how the big picture guides one executive in his chance hallway meetings with staff. On his way to a meeting, the executive bumps into a staff member who does not report to him. During a two-minute "hallway meeting," he asks the staffer several questions, pays a compliment about a job recently well-done, and gets the staff member to agree to do something that he needs. To set up and attend a meeting to accomplish the same outcomes, the executive would have had to spend at least 15 to 30 minutes, or 750 to 1,500 percent more time than the chance encounter took. By keeping the big picture always in mind, the executive can turn chance encounters into productive sessions.

Some years ago, I reported to a supervisor who believed in two-minute hallway meetings. It's amazing how much can be taken care of that way. The group rarely had occasion to meet, because issues were handled as they arose. We spent our time getting things accomplished rather than sitting in boring meetings.

Time Management Tools

Traditional time management tools such as time planners and calendars can actually undermine effectiveness because of their emphasis on daily planning. One manager in a Fortune 500 company discovered this when she began using a popular planning system. She found it frustrating, because she wasted time writing down and keeping track of numerous bits of information while nothing important got accomplished.

To be helpful, a time management tool must be converted to a Master Control System as described in the previous chapter. Then you can view details in the **context of priorities.** Avoiding overload requires a perspective on the total situation. Without this perspective, information is treated as isolated bits instead of as pieces of a unified whole. The value of the MCS depends on whether you use it to:

- View the big picture.
- Track details in light of priorities.
- See how parts relate to the whole.
- Integrate with other systems.

One time management system, the Franklin Day Planner, teaches users to relate details to personal life goals by visualizing a

"Productivity Pyramid." With this approach, the planning process begins as you define your basic values as the foundation of the pyramid. You then layer intermediate goals and daily tasks on this foundation so that everything you do relates to your personal values.

Values form the foundation of the "Productivity Pyramid."
Source: Franklin International. Used with permission.

"I cannot tell you what an impact the pyramid of values has had on my life," says one enthusiast of the pyramid. "I know where I'm going and why." Another person sees the value of the pyramid as helping "put life's activities into perspective."

To optimize your effectiveness, you'll need to expand your perspective to see your work activities in a broad context. That perspective will allow you to distinguish between activities that contribute to your effectiveness and those that should be delegated or eliminated.

Expanding Your Perspective

It's vital to envision in your mind's eye what you're aiming for. The point is to keep the big picture in mind as you plan and execute your daily, weekly, monthly, and even yearly activities.

Expanding your perspective is a matter of reviewing in your mind what your work is all about. How does each activity relate to your major goals and priorities? What tasks really matter in your department, your company, your industry? You should give highest priority to activities that:

- Matter most to your department, company, industry, and customers; *and*
- Follow from your vision and contribute most toward the attainment of your main goals.

The External Context

Expand your perspective to see yourself and your job in *context*. Your external context includes the conditions and circumstances in your department, company, and industry. Which activities will be rewarded by your organization? It also encompasses your customers' needs and wants as well as those of your suppliers. These influences often seem to be at odds: the company wants you to cut costs and save time, the supplier wants to sell you more goods or services, and the customer wants a better product or service at a lower price. As a professor, I continuously grapple with student demands to spend more time teaching, the organization's dictum to "publish or perish," and appeals from textbook and software suppliers to buy from them.

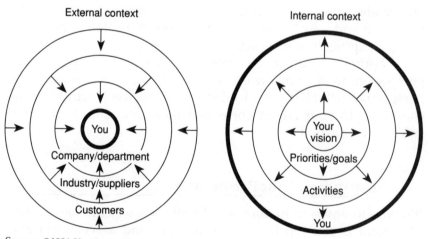

Source: ©1991 K. Alesandrini

It's risky to ignore the external context as you decide how to allocate your time and resources. The classic case of ignoring external context can be summed up in one word: Edsel. Ford designed the car with no consumer research to find out what customers wanted, named it after Henry Ford II's father, Edsel, and wasted an estimated $350 million in the process on tooling and production costs alone. Nearly a decade later, Ford introduced one of its biggest success stories after extensive consumer testing and followed an ad agency's advice to name it: the Mustang.

Your Internal Context

It's important to envision the internal as well as the external context. What is your overriding vision? What are your motivations, skills, and experience? These elements shape your priorities and goals. It all begins with vision:

Vision ⟶ Priorities ⟶ Activities

What if your internal and external contexts lead you to irreconcilably different priorities? You could try to change the external context (i.e., persuade your boss to alter your job objectives), seek a new external context (i.e., change jobs), or adjust your internal context (i.e., abandon your own goals). For anyone who'd like to be happy, the latter alternative is not an option.

It is often necessary to change the external context. Its influence on you is no one-way street; you can change your boss's mind, your department can modify company policy, the company can sway consumer opinion. American business spends over $130 billion annually in advertising to try to sway, cajole, persuade, inform, manipulate, or otherwise change the external context that consumers represent.

Are Companies Paying Attention to Context?

Has business learned its lesson about paying attention to customers and suppliers in the 1990s? Not all that well, according to a survey of nearly 12,000 managers in 25 countries conducted by the

Harvard Business Review. Customer service is seen as the top success factor in nearly every country. Yet in practice, "customers are kept at arm's length and suppliers only a little closer." When asked, "Does your organization share strategic information with customers?" nearly one fourth—23 percent—said "never." Two thirds of the managers reported the frequency as "sometimes" or "never." Less than one third included customers on a product-planning team on a frequent basis. Evidently, managers are not practicing what they preach about the importance of the customer.

Suppliers tend to be brought into the loop more frequently than do customers. Over 75 percent of the companies reported long-term relationships with suppliers. Most companies, however, do not involve suppliers in product design on a frequent basis.

How can companies involve customers, suppliers, and other "outsiders" to a greater degree? Here's what some are doing:

- Focus-group research with prospects and customers.
- "Outsiders" serving as directors on company boards.
- Reorganizing the company from a "functional" to a customer-led market organization.
- Adding more product and sales groups to respond to customer needs.
- Involving customers and suppliers in product development efforts.
- Bringing in outside trainers and consultants with "new eyes" and fresh ideas.
- Recruiting customers and suppliers to get involved with customer service.

Wal-Mart Stores, voted number one in quality of management, serves as an example of paying attention to context. The company's CEO, David Glass, spends at least two full days weekly visiting his stores and his competitors' outlets. He asks Wal-Mart clerks and customers for ideas and complaints. "I can chart the course for the company," he says, "because I get exposure to ideas that the rest of our employees don't."

The external context is important, yet perhaps what is even more important lies at the core of the internal context.

The Power of Vision

The thing that matters more to success than just about any other factor is the vision you have of what your project, division—or entire company—could become. And the more you think about the vision's meaning to yourself and to the organization, the better you'll be at inspiring others to work at making the vision a reality.

It's important to let your vision possess you, because your passion gives the vision credibility. "I have a one-track mind," admits Harley-Davidson's vice president of marketing, Kathleen Lawler-Demitros. "Whatever the vision is, it pervades everything that I do. . . . I'm constantly communicating it so that it becomes the thing that all people around me focus on. That's the key to success."

Her vision was to celebrate the company's 85th anniversary with a cross-country motorcycle marathon that would also benefit charity. With total dedication to her vision and eight months work to back it up, she pulled off the event, which was the largest in the company's history. Corporate officers led 35,000 Harley customers from 10 cities around the United States to a festival in Milwaukee. The event not only raised a half-million dollars for charity, it earned an award as the best PR event that year. "Through constant visualizing of the idea and the magnitude of it," says Lawler-Demitros, "I made this ride unique."

Communicating Your Vision

For your vision to be useful, it must be communicated. Most top executives believe in the value of communication with their employees, but few actually do it. Here's proof:

- 97 percent of CEOs surveyed by A. Foster Higgins, an employee-benefits consulting firm, believe communicating with employees aids job satisfaction and 79 percent think it benefits the bottom line, but only 22 percent actually do it weekly or more often.
- Employees want to know "where the company is headed" and "how my job fits into the total" but aren't being told, according to research by the Hay Group.

- Less than one third of employees polled by Louis Harris said management provides clear goals and direction.
- Yet 82 percent of Fortune 500 CEOs surveyed by the Forum Corporation said their strategy was "clearly understood by everyone who needs to know." Fewer COOs and CFOs agreed (68 percent and 62 percent respectively).

Evidently, there is a discrepancy between how well managers think they communicate their visions and how well the message is getting across. Sometimes the easiest way to communicate your vision is to simply sit down and tell others about it. The head of an insurance company once sought out a consultant with the complaint that employees were not "getting" his corporate vision. He had sent out some materials and made a video. When asked if he had met with them face-to-face, the CEO was shocked: "You mean, sit with them in little red plastic chairs and drink coffee out of Styrofoam cups?" He had not done that.

Other approaches may be needed to make the vision relevant to others. "Look for the salient images and phrases," advises a magazine editor, "that not only capture your vision most forcefully but make it accessible to others." He suggests you keep a notebook handy to keep track of your best one-liners, selling points, and the critical needs that your vision will fulfill. "Look especially for metaphors, analogies, and anecdotes that you can use to bridge the gap between your vision and the grasp others have of it."

"You must communicate not only your excitement," says Joan Spero, senior vice president at American Express, "but also the significance of what you're doing—why it matters to the company and why the mission you have *means* something." Responsible for positioning American Express in the European financial-services market, Spero has communicated her vision of what is at stake: "If American Express has access to the new and growing European market, it could be a huge business opportunity."

There's another, compelling reason to share your vision with employees. Communicate with your staff, and they, in turn, communicate with the customer. Robert Kelley, who teaches management at Carnegie-Mellon, observes, "Service providers treat customers similarly to the way they, as employees, are treated by management." Ultimately, your vision will impact the customer.

Why Visualize?

"Whatever you materialize in your life must first be pictured in your mind by your imagination," advises cookie magnate Wally Amos, who credits his rise to riches largely to vision. "It really is possible to take an idea right out of your mind and make it become real." The mental vision of what you want to accomplish fuels your efforts to pursue your priorities.

While a mental vision is good, committing the vision to paper is even better. Maps, charts, and other graphics depict the vision in a way that can be communicated directly. Although I should have realized that *seeing* the big picture is better than mentally thinking or reading about it, I had to re-learn this truism for myself while writing this book.

Many writers use an outline as a guide while they write. A written outline, however, does not necessarily convey the big picture. I relied exclusively on a mental vision and a written outline for many months as I did the preliminary research for this book. One day I showed the outline to a colleague. She pointed out that several sections did not seem to fit very well with the overall theme of how to *survive information overload*. For example, there was a section on how to design visual aids for an effective presentation. At first, I resisted her comments. I told her I would create a simple chart that graphically showed how the section fit into the bigger picture.

When I finally sat down to visualize the big picture, I discovered that my colleague was right. I had gone off on a tangent (as we professors are wont to do). In addition, several sections needed to be rewritten to address the issue of information overload more directly. Mapping the big picture for the book took less than 20 minutes and saved untold hours of needless work. Then as the project evolved further, the map was simplified to a seven-slice pie corresponding to the letters in the word S-U-R-V-I-V-E.

The Need to Prioritize

Richard Bolles, author of *What Color Is Your Parachute*, recommends: "I think almost everybody today has some problem with time. They are never, ever going to get done all that they want

Box 2–1

Prioritizing: How Executives See It

• Uses a personal information manager (PIM) to assign activities to major categories. "I can sort by categories, dates, people, and priorities, depending on what I need to know at that moment."
James Treleaven, *venture capitalist*

• Focuses on goals, maintains an awareness of strategic position, strengths, and weaknesses. "We don't waste a lot of time debating businesses that don't fit our strategy."
Anthony Burns, *CEO, Ryder System*

• Decides ahead of time what proportion of his time he wants to devote to certain activities, then allocates specific blocks up to a year in advance.
Barry Sullivan, *CEO, First Chicago*

• Practices "just-in-time" worrying: not bothering with a task until he absolutely must. "If you worry too soon, things will change in the interim, so you end up having to deal with them twice."
John Young, *CEO, Hewlett-Packard*

• Delegates details so she can focus on strategy. "I found my strength lay less in details than in my ability to . . . see the big picture."
Sandra Kurtzig, *CEO, ASK Computer Systems*

• "You either commit to something 99 and 44/100ths percent or you don't bother with it at all."
Allen Rosenshine, *CEO, BBDO Worldwide*

to do and therefore they have to establish priorities. They have to get a vision in their head of what is most important to address."

Once you have your priorities in mind, you can set clear goals. According to a recent survey of 439 managers by *Industry Week*, not all managers have clear goals. "Of top management, 80 percent answer 'yes' to clear job goals. Not bad. But only 70 percent of middle management and 60 percent of first-level management say 'yes' to the same question." Issues to address include:

* Which goals are most critical?
* How does the context affect priorities?
* Which activities have a high priority?
* How are tasks related to one another?
* Which activities could sidetrack efforts to pursue priorities?

Prioritizing should be an ongoing process. It's not enough to set your priorities once, because the context is always changing. Rod Canion, founder and former CEO of Compaq Computer, believes in the value of continually reprioritizing. When he first started the company, "I knew what everyone else was doing every hour of the day," he says. "But you can't do that with 9,000 employees." He has had to reassess repeatedly what he felt was the most important area to spend his time in. "You have to prioritize your activities again and again—otherwise something is going to snap."

Like other effective executives, Canion looked beyond the urgent demands and spent time thinking about the big picture—the future of his firm and how to counter the competition.

Prioritizing for Effectiveness

Prioritizing increases your effectiveness as you spend time on the right things and avoid time wasters. Time management is concerned with achieving *efficiency*. Seeing the big picture and setting priorities help you achieve *effectiveness*. The following lists illustrate the difference between the two.

Effectiveness	*Efficiency*
Setting priorities	Setting standards
Meeting objectives	Meeting deadlines
Achieving major goals	Achieving results
Doing the right things	Doing things right
Significant	Competent
Vision	Time management

The challenge of today is to develop a vision of your priorities. Dr. Stephen Covey, chairman of the Covey Leadership Center, observed that successful people "begin with the end in mind." He calls this habit end-result thinking and sees it as essential for effectiveness. He believes we will "be truly *effective* only when we begin with the end in mind."

End-result thinking means paying attention to where you're headed and looking ahead to see if you want to end up where your current direction is leading you. The end result to a great extent will determine your current direction once you become a big-picture thinker.

Deciding What NOT to Do

The more valuable your time, the more important it is for you to decide what *not* to do, not what *to do*. When asked what type of music he liked, Mozart replied that he actually preferred silence—"rests" in musical jargon. His passion was "music-accentuated *silence*." To him, what he left out of his music was more important than what he put in. In a similar vein, the 19th-century painter James Whistler defined success as knowing what *not* to put on canvas. To escape information overload, you must decide what *not* to do in order to have the time to pursue priorities.

One useful tool for figuring out what to leave out is the time management matrix. The matrix is one time management concept that remains useful in the information age. All of your activities can be classified into the matrix on the basis of their importance and their urgency. Some activities are both important and urgent, like the imminent deadline of a major unfinished project. Too often, however, we ignore important activities to cope with urgent

Time management matrix

	Urgent	Not urgent
Important	I • Crisis • Imminent project deadline • Emergency • Pressing problem	II • Learning, seeing the big picture • Exercise, recreation, relaxation • Networking, relationship building • Prospecting, finding new opportunities
Not important	III • Some phone calls • Some meetings • Some mail, some projects • Some interruptions	IV • Busywork • Some television viewing • Some phone calls • Time wasters

ones. A knock at the door and a ringing phone urgently demand a response, but the matters may be trivial. Studies confirm that most urgent matters are not important.

The biggest problem with urgent matters is that they blind us to the big picture. We can't see the forest for the burning tree right in front of our eyes. You could miss a conflagration developing in another part of the forest while you spend time putting out the small blaze that someone else can more easily put out or that might burn itself out. Urgent matters demand attention and immediate action. If you've waited until a matter becomes urgent, it may be too late. "Many of the crises that arise in business or in personal life result from failure to act until a matter becomes urgent," warns the author of *Getting Things Done*, "with a result that more time is required to do the job." You'll end up wasting time with typing or copying at the last minute because your staff had no lead time.

When you are overloaded with urgent matters, there is no time for activities that are important but not urgent. You pay a price— it's called opportunity cost—for being too busy to do what's important and to take advantage of an opportunity when it knocks. Without a firm vision of your priorities, you can easily get involved in projects and busywork that not only waste time but also prevent you from taking advantage of the right activities—those that move you closer to your goals.

The first step of the following Action Plan helps you sort activities according to what should become your priorities and what you can eliminate.

ACTION PLAN: Underscore Your Priorities

Underscoring your priorities is a four-step process:

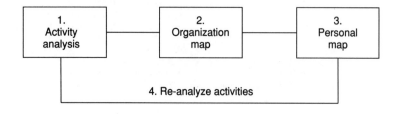

Step 1—Conduct an Activity Analysis

Spend a few minutes mentally classifying a dozen or more of your most time-consuming activities from a recent typical workday into the four "Quadrants" of the time management matrix.

Time management matrix

	Urgent	Not urgent
Important	I • • • •	II • • • •
Not important	III • • • •	IV • • • •

Are you consumed by major crises in Quadrant I? If much of your time is consumed by urgent and important matters such as imminent project deadlines and problems that need immediate attention, you will soon be a candidate for stress and burnout if you haven't reached that point already. Everyone has occasional

major crises and emergencies, but an abundance of pressing problems is a clear sign of misplaced priorities. Begin now to minimize
Quadrant I activities:

* Ask yourself: "Is this matter truly important and urgent?"
  Don't assume that because something is urgent, it is also
  important.
* Extend your short-term focus to a longer-term view.
* See goals and plans as valuable tools to free you from a
  crisis mentality.

Make important activities your priorities. If you were
able to think of only a few activities for Quadrant II, take a few
moments now to add important activities that you would like to do
if only you could find the time. Ask yourself what things you could
do that would yield an enormous positive improvement in your
business or career. Candidates include learning, developing and
maintaining vision, networking, managing important projects,
and so on. These activities represent the core of how you should
spend your time—putting first things first. You'll learn how to
map your important activities in the subsequent steps of this
Action Plan. Here are two basics:

* Plan on a weekly rather than daily basis, since daily planning
tends to focus attention on urgency rather than importance. You'll
still adapt on a daily basis, but the point is to organize within a
broader framework to maintain a big-picture view. Effectiveness
expert Stephen Covey advises, "The key is not to prioritize what's
on your schedule, but to schedule your priorities. And this can
best be done in the context of the week."

* Include both professional and personal activities when planning your priorities. After all, you don't want to neglect important
personal matters such as exercise, relaxation, and personal relationships. Make sure that they're included in your big picture.

Delegate or drop unimportant activities, urgent or not.
Initially, *making* time for important matters means *taking* time from
busywork and time wasters. You cannot ignore the urgent and
important matters of Quadrant I, although this category will di-

minish as you do a better job of planning and prioritizing. Quadrants III and IV represent opportunity cost. You cannot spend your time doing important tasks if your time is spent on unimportant matters. The goal, then, is to delegate or eliminate these activities altogether.

- Ask yourself if the urgency of Quadrant III activities is based on your own priorities or on the expectations and priorities of others.
- Learn to say "no" pleasantly but firmly to activities inconsistent with your big picture.
- Delegate Quadrant III activities to catch anything that may be important after all. Communicate *what* needs to be done, not *how* it should be accomplished. Focus on results, not methods.
- Provide or identify adequate resources when you delegate so the person has the means to accomplish the desired results.

Step 2—Map Your Organization's Big Picture

The *process* of mapping the big picture is more important than the aesthetics of the resultant *product*. Your map will only be useful inasmuch as it stimulates you to see the big picture during your daily and weekly activities.

Whether you spend 10 minutes or 10 hours on this activity, you won't find a better investment of your time. If you have more than 10 minutes available now, begin by mapping your organizational context. Otherwise, skip to the next step and map your individual big picture.

You can accomplish this task with the simplest tools: pencil and paper. Or you can use specialized charting software (see "Outlining & Diagraming Software" in the resource directory). Using this type of software, you select symbols such as boxes and then choose from a variety of different ways to route connecting lines. If you move a box, the lines automatically reroute. To simplify the task of chart creation, some programs produce charts automatically from a written outline.

One useful tool for creating organization maps is *TopDown*. It provides 10 different ways to route lines automatically.

Source: Illustration courtesy of Kaetron software.

Create a basic organization map. One way to maintain a big-picture perspective is to map your organization on paper. A valuable resource is Carol Panza's booklet, *Picture This . . . Your Function, Your Company.* Mapping your organization on paper will help you put your own responsibilities into perspective.

A traditional organization chart of your company shows how it's organized but says little of how it *works.* In contrast, an organization map is a graphic representation of how an organization works. It shows the performance context in which the company operates, something the people of an organization need to understand for it to survive long-term.

The figure on page 50 is a generic example of a simplified organization map. It forces you to take a big-picture perspective on your organization; suppliers are included since they provide the organization with certain essentials such as raw materials, labor, and energy. The organization converts this input into products or services through a number of processes or structures for some customer or client.

Notice the two feedback loops. The first is for management controls based on what the organization thinks is good. The other is for feedback from the market about their demands and how well the organization is meeting their needs.

Carol Panza, author of the mapping guide, advises, "You shouldn't just 'map it' and forget it. The lesson here is that you must define your organization's context AND you must keep that

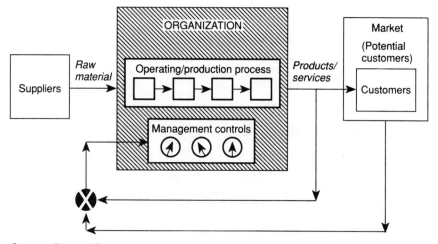

Source: *Picture This . . . Your Function, Your Company.*
Source: © 1991 by Carol M. Panza.

'picture' up-to-date. If you do 'map it,' it is in your best interest to use that picture as a basic framework to periodically ask questions about what surrounds your organization. The external environment, the world outside our organizational doors is dynamic." It's a good idea to continually assess the impact of changes in market needs and wants, local and global competition, government regulation, technology, suppliers, labor, and so on.

Create a relationship or cross-functional map. A relationship map helps individuals in one unit see themselves in relation to other units and to the organization as a whole. The mapping process often exposes inconsistent or disjointed efforts.

Suppose, for example, the sales department projects a 20 percent growth rate and focuses on that objective. At the same time, customer service tries to hold costs to no more than 5 percent above last year. Each group's objectives seem reasonable from within their own departments. However, a relationship map reveals conflicts that are not evident until you look across functions and attempt to coordinate each group's objectives in support of organizational goals.

A relationship map can be used to clarify the most complex organizations. A pharmaceutical company, for example, mapped its entire business: the development, manufacture, sale, and dis-

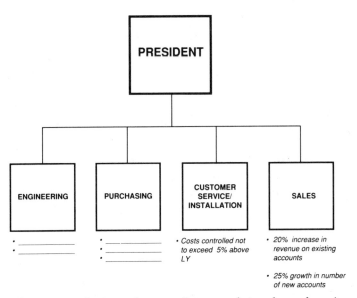

A big-picture view reveals contradictory goals in sales and service.
Source: © 1989 by Carol M. Panza.

tribution of its products. Although the original purpose of the mapping project was to provide an orientation map for new hires in the rapidly growing company, the map helped middle and senior managers better understand how the functional parts of the company interrelated.

The result of the effort was cooperation and multifunctional teaming to solve problems. Carol Panza, who directed the mapping effort, told me the activity was the "best kind of team building that could possibly happen. People were really enthusiastic about an opportunity to work with their colleagues in a frame of what needs to get done. . . . Where there had once been 'fights to the death' over the power base issues bound up with organization chart development and revision, there was now a realization that 'we're all in this together and none of us will truly be successful unless we are all successful.' "

Create a process map. A training group at Motorola mapped the process for scheduling training throughout the organization. Scheduling was handled at several sites, and the

resultant 32-step map revealed several discontinuities and duplication of effort. The task was reduced to a simple, four-step procedure that was automated, an 88 percent reduction.

A process map documents and clarifies business processes of
individual departments, projects, or activities. One organization,
a telephone company, used a process map to improve its customer
payment process. It created a task force from different departments to map the process. Task force members worked together,
each contributing expertise on how their function handles the
payment process. The resultant map was used by management to
see how the key parts of the payment process fit together so that
better management controls could be developed. Before the
process map, the big picture had never been articulated.

A process map can be used to clarify or train an employee to do a
process. The following example shows a simple map of the telephone answering procedure at a software company. Notice that
the verbal description is linear: do this, then do this, then do this,
and so on. It buries the big picture in verbiage. The reader could
easily miss the fact that there are three basic types of incoming
phone calls. In contrast, the map immediately conveys the three
categories and shows how to respond appropriately to each.

Step 3—Map Your Individual Big Picture

Starting with the dozen or so categories of responsibility that you
identified during Step 1 of the Action Plan in the previous chapter,
you'll create a chart that provides a simplified overview of your
main activities and priorities.

- Use a box to represent each of the dozen or so most
  important areas of your work. Take a look at the categories
  of responsibility you identified in the previous chapter.
  Revise them as needed based on the results of Step 2 of this
  Action Plan. For example, now you may see customers as
  an important category to delineate. (Use a small self-stick
  note for each box if you are going "low-tech.")
- Title each box with a one-word category description such as
  "Prospecting." Add all boxes to a sheet of paper or screen
  page in any arrangement.

Telephone Answering Procedure

Answer the phone "Good morning, Kaetron Software."
It is not necessary to give your name when you first answer the phone.

• Is the person calling with an Inquiry?
If yes:
Get name & phone number.
Say "I will transfer you to someone in Sales."

If not:
• Is the person calling with a Problem?

If yes:
Get Name, phone number and serial number.
The serial number for version 1.x is in the inside front cover of the manual. These serial numbers are 10000-14000.

The serial number of 2.x and 3.x is in the "About Box" These serial numbers are: FX000000 or HSX000000.

Say "I will transfer you to someone in Customer Support."

If not:
• Is the person calling with an Upgrade? If yes:
Get name & phone number
Say "I will transfer you to someone in Sales"

Is someone available?
First try anyone in the department, if no one is available, then transfer call to the department manager. Never say I'll transfer you to the owner, sales girl

If so:
Give memo to the person (mark line number.)

If not:
Ask if you can take a message.

Don't ask what the problem is or what it is in regards to. If they need someone right away, they will tell you.

Telephone answering procedure

Answer the phone "Good morning, Kaetron Software." — It is not necessary to give your name when you first answer the phone.

Upgrade · **??? Inquiry** · **Problem**

Get name and phone number

Get name, phone number and serial number.
- The serial number for version 1.x is in the inside front cover of the manual. These serial numbers are 10000-14000.
- The serial number of 2.x and 3.x is in "Get Info" box and the "About Box." These serial numbers are: F000000 for version 2.x and HS000000 for version 3.x

Say "I will transfer you to someone in Sales."

Say "I will transfer you to someone in Customer Support."
- Never say "I'll transfer you to the owner (sales girl)."

Is someone available? — No / Yes

Ask if you can take a message. — Don't ask what the problem is or what it is in regard to. If they need someone right away, they will tell you.

Give memo to the person (mark line number on it.) — First try anyone in the department; if no one is available, then transfer call to the department manager.

Source: Examples courtesy of Kaetron Software.

- Think of two or three key results in each area that you would like to accomplish during the next year. Use your list of Quadrant II activities from Step 1 as a guide. These results represent your intermediate-range goals.

```
┌─────────────────────────────────┐
│  Products   Services    People  │
│                                 │
│                                 │
│                                 │      ┌──────────────────────────┐
│                                 │      │      Prospecting         │
│                                 │      │                          │
│                                 │      │  • Track 500 leads       │
│                                 │      │  • Computerize follow-up │
│                                 │      │  • Close 15% of leads    │
│                                 │      │                          │
│                                 │      └──────────────────────────┘
│                                 │
│                                 │
│                                 │
│                                 │
└─────────────────────────────────┘
```

- Describe each goal in a few words and write key goals on each of the boxes.
- Identify several general, inclusive headings under which all other categories (the boxes) can be organized such as Products, Services, and People. Write these headings near the top of the sheet from left to right.
- Sort the boxes into vertically aligned groups under the inclusive headings that you identified in the previous step.
- Within each large group, order the boxes vertically based on their importance with the most important on top. If several activities are related or equally important, show that relationship visually by placing them at the same height.
- Add connecting lines to indicate relationships both within and between groups.
- Repeat the process for each individual area of responsibility to create a big picture for that area or project. In other words, create a breakout map for each of the boxes in the top-level map you have just made. Flowcharting software allows you to create "embedded" maps that can be accessed on a new screen display.

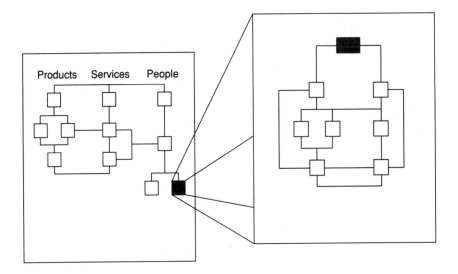

This is your individual big picture for the next year and should be placed in the overview section of your MCS. If you make duplicates of this layout without the goal phrases, you can fill in weekly goals and place each weekly view next to your daily calendar page. Carry both views in your MCS to keep an eye on both the weekly and the yearly big picture.

Step 4—Mentally Redo Step 1 in Light of Steps 2–3

Go back and take a look at the time management matrix that you created for Step 1 that sorts your activities by importance and by urgency. Once you have carefully considered the external context and mapped your priorities in the previous two steps, your categorization will have greater validity.

- Were activities dropped that could provide important information about your firm's external context? These activities might include chatting with customers or reading about your suppliers.
- What is the main source of urgency—dissatisfied customers, suppliers who don't deliver what they promise, last-minute pressures to finish important projects that you put off? Your answer may signal needed changes in either the internal or external context rather than your activities.

• Do you now see inconsistencies between what you classified as important and what the context says is important? If so, you may need to adjust your priorities.

Generally, when people go back and redo the time management matrix after gaining a big-picture perspective, the activity triggers lasting changes in how they spend their time.

SELF-TEST: How Well Do You Manage Your Time?

	Yes	*No*
1. When you arrive at the office, do you have a clear vision of your primary goals and priorities?	____	____
2. Do you generally accomplish your top priorities on a weekly basis?	____	____
3. Do you feel you have more pressing problems, crises, or emergencies in your work than most people?	____	____
4. Do you feel compelled to read everything that comes across your desk?	____	____
5. Do you avoid reading or even looking at information outside your specialty, because you assume it has no relevance to your business or personal life?	____	____
6. Do you have a one-page outline, map, or chart for the most important activity you are currently working on?	____	____
7. Do you have a few clear goals for each of your projects or activities?	____	____
8. In the last month, did you meet with an "outsider" because he or she had insights for your business?	____	____
9. In the last three months, did you end up copying or typing at the last minute because your staff had no lead time?	____	____

10. In the last three months, did you put off
a project until it required a panic of
overtime work to complete?   ____   ____

Scoring

Questions 1–2, 6–8: 1 point for each "no."
Questions 3–5, 9–10: 1 point for each "yes."

What Does Your Score Mean?

Score of 0   Good for you! You are accomplishing what's
important in the time available. Skim the
chapter for any new ideas and go on to the
next chapter.

Score of 1–3   You are managing your time fairly well now
and can achieve peak performance by reading
through the Action Plan in this chapter.

Score of 4–6   You can increase your productivity
substantially by studying this chapter and
working through the Action Plan.

Score of 7–10   Ineffective strategies compromise your
productivity. Make sure you study this
chapter thoroughly and follow the Action
Plan carefully. Invest extra time now and reap
the dividends of new techniques for
controlling your time and your life.

If you still have "too much to do, too little time," the problem
may be "too much paperwork." According to a survey of 414
businesses, paperwork ranks as one of the top five complaints
along with costs, cash flow, and employee relations. It is essential
in the information age to learn how to cope with paperwork. Turn
to the next chapter to find out how to "reduce the excess."

Chapter Three

Too Much Paperwork? . . .
Reduce the Excess
(with Electronic Paper, Color Coding, and Visual Organization)

In today's office, your choices are two: Manage paper or it will manage you.

It's impossible to find time to read all of the publications, memos, reports, and correspondence coming across your desk. Chances are you have 35 to 42 hours of paperwork within arm's reach of your desk chair but only 90 minutes to spend on it, according to industry estimates. Yet not keeping up with paperwork can be a serious and expensive mistake. You could miss important ideas, opportunities, and insights.

The pace of change has accelerated more in the past decade than at any other time in history. To succeed in the face of rapid change, you *must* keep up in order to respond effectively to competitive, economic, and technological developments. To achieve a high level of productivity and success, conquer the paperwork challenge by reducing the excess with "electronic paper," color coding, and visual organization.

A TON of Paperwork *Does* Cross Your Desk

If it seems that you are drowning under a ton of paperwork, you probably are—literally. The total weight of paperwork crossing the average manager's desk in the course of a year approaches the one ton mark! Check out the numbers:

- We now use 22.8 million tons of office paper annually, which amounts to over one ton for every executive, administrator, and manager in the work force.
- Demand for office paper has grown twice as fast as the gross national product for the last 10 years.
- Every day the United States generates 750 million pages of computer printout, 234 million photocopies, 76 million letters, and 21 million other documents.
- Over 60 billion pieces of third-class or junk mail weighing 4 million tons are sent each year, amounting to 41 pounds for each adult American.
- In total, roughly 1.6 trillion pieces of paper circulate through the nation's offices each year.

Paperwork accounts for nearly half of the white-collar payroll in the United States and costs $100 billion annually, not including the cost of filling out government forms, which adds an additional $166 billion annually. In an attempt to ease this burden, the Paperwork Reduction Act lopped a significant yet insufficient 500 million hours off the billions spent each year on paperwork. Depending on whose statistics you believe, white-collar workers spend anywhere from 20 to 70 percent of their working hours on paperwork. By any measure, paperwork is out of control.

Who Faces the Most Paperwork?

In a 1982 landmark study of how workers spend their time, several hundred white-collar workers kept records of their daily activities. Results show that these workers, on the average, spent 21 percent of their time reading and writing.

The percentages may be even higher, according to a report by Facts on File of 657 white-collar workers in 14 companies from seven different industries. Nearly half of the average worker's day was spent dealing with paperwork. According to the breakdown by job category, managers spend about the same amount of time on reading as they do on writing. Middle managers, for example, spend 23 percent of their time reading and 22 percent writing. In contrast, professionals spend more time writing (22 percent) than reading (18 percent).

The paperwork burden is particularly heavy for some:

- Paperwork accounts for 24 cents out of every dollar spent on health care in the United States.
- Each medical insurance claim consumes six minutes of a doctor's time and an hour of secretarial work.
- 37 percent of small-business owners rate paperwork or governmental regulations as their top complaint.
- The 535 members of Congress face 3,000 recurring reports plus thousands of one-time reports each year, despite passing the Congressional Reports Elimination Act several times in the last decade.

No Need to Shuffle "Electronic Paper"

The best way to handle paperwork is to eliminate the need for it altogether. You can do that by converting it into "electronic paper." The only good reason for using paper these days is to *display* information in a form that you can read easily. To *store* or *transmit* information, however, paper is a lousy medium, because it's too cumbersome. Electronic media are far superior to paper and generally less costly besides.

To illustrate the superiority of electronic media, consider the case of an electronic database containing periodicals. *Magazine Index,* for example, contains the full text from over 530 popular magazines. The charge to search for topics of interest is $1.50 per minute. You must also pay for each article that you retrieve, often bringing the total cost to several dollars per article. You can get the information you need at less than you would pay if you subscribed to just a dozen or so of these publications. When you add to the cost of traditional magazine subscriptions what you "pay" in reading and searching time to find relevant articles, electronic media look all the more attractive. With paper-based media, you could miss vital information altogether. What if your competitor was mentioned in *Newsweek* but you read *Time?* You could miss a cover story in one of the 288 magazines you don't read that would change everyone's perspective in next week's meeting. With the electronic medium, you zip right to the article on your competitor, pinpoint a needed statistic, or check press coverage of your company within seconds rather than hours.

The aim is not for a "paperless office," because that worn-out notion misses the point. Today's office needs a variety of information media, including paper. It's time white-collar workers put paper in its place and turned to electronic media for many, if not most, information needs.

Fax-It-to-Me

As a means of *transmitting* paper electronically, the facsimile machine has taken the business world by storm. According to a recent survey by the International Data Corporation, 99 percent of Fortune 500 companies use fax machines. An estimated 60 billion faxes will be sent this year if you count all the faxes sent by computer. The numbers are mounting:

- 5 million fax machines in the United States.
- 15 million worldwide.
- An average of 12 pages per day are sent by each stand-alone fax machine.
- The number of fax messages sent by 1995 will be up 250 percent over 1990.
- The computer fax market is doubling each year.
- A single major investment bank sends 1 million faxes a year by a computer programmed to dial hundreds of phone numbers automatically.

When it is used well, the fax machine can keep you from getting bogged down. Marilyn Ounjian, CEO of Careers USA, turned to daily faxes to keep tabs on a sales force spread over 21 offices in nine states. It was difficult to stay on top of what was happening since she visited some offices only twice a year. In order to regain management control over the dispersed sales force without a barrage of paper, she came up with the idea of using a one-page sales summary, faxed to her daily.

"Simplicity is key," explains Ounjian. "I don't ask for anything on this sheet that's not absolutely necessary." She combs the summaries for trends and discrepancies. "Financial statements tell me the past, but there's not much I can do about that," she says. "I want to know what's happening now." Although the information is eventually entered into a computer, the handwritten reports sent via fax keep her current.

Fax for Better Customer Service

Some companies are using facsimile machines as a customer service tool. Borland, for example, provides its customers with an automated voice-fax system called "TechFax." The customer selects needed information from a menu of choices presented over the phone. Then the system automatically faxes the information to the customer's fax machine. When I asked customer support manager Jim Fitzgerald how it was working out, he explained, "It's cheaper than having customers talk with an engineer. We're right next to the heart of 'Silicon Valley,' and it's very expensive to do support." The main purpose of the system, however, is not to save money or paper. "The main purpose is to provide an enhanced level of service to the customer."

I was so impressed with the automated information-on-demand that we are instituting a similar setup. When organizations call us to inquire about our seminars or consulting services, they'll receive immediate answers via facsimile. The arrangement allows us to maintain a lean support staff while providing timely information to our clients. The technology supporting this service is called the FaxPump (so named because it pumps out faxes automatically).

Electronic Mail

Another alternative to paper transmission, electronic mail (E-mail) is no longer simply a way to exchange personal messages via computer. E-mail is used by 12.8 million people to send 1.2 billion messages annually, addressing a variety of business needs both within and between organizations. Some 67 percent of Fortune 500 companies report using some type of E-mail system. The growth

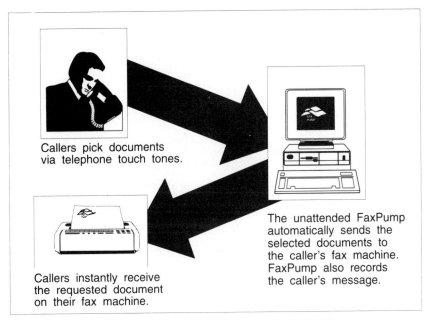

Callers pick documents
via telephone touch tones.

The unattended FaxPump
automatically sends the
selected documents to
the caller's fax machine.
FaxPump also records
the caller's message.

Callers instantly receive
the requested document
on their fax machine.

©1991 FaxPump, 181 N. Central Avenue, Campbell, CA 95008.

has been phenomenal, up nearly 3,000 percent in the last 10 years. Growth of E-mail at DuPont is a good example. They went from 70 users to 90,000 employees who send about a half-million messages worldwide *every day*. The rate of usage is growing by more than 60 percent per year.

Until recently, there were generally no links between E-mail systems. You could send only to other users on the same system. The four million users of IBM's PROFS, the most popular system, for example, communicated among themselves. Now technology that links systems can bridge the gap between the many "communication islands" such as ALL-IN-1 from Digital, AT&TMail, cc:Mail, CompuServe, MCI Mail, Wang OFFICE, and others.

Will E-Mail Ease Your Workload or Add to It?

Not everyone sees electronic mail as a way to lessen information overload. Sometimes it opens a new channel of communication that quickly becomes saturated. A well-known speaker told me that he stopped using a popular E-mail system because "I ended

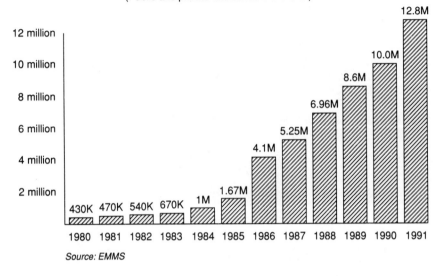

The growth of electronic mail
(Public and private mailboxes in the U.S.)

Source: EMMS

up getting too much to answer. It was taking too much of my time just to go through it." Most of his E-mail was from people in his audiences who had questions or comments. "I ended up getting swamped. You don't realize until you open the channels of communication how many people want to be involved."

A software engineer at Digital Equipment Corporation was also swamped with E-mail. Returning from a three-week stint out of the office, he was horrified to find 1,000 electronic messages waiting for him. Since each message has a code that identifies the sender, he programmed his computer to prioritize messages according to who wrote them. Now E-mail from his supervisor goes into a high-priority file, while routine announcements are put on the back burner. Messages from some senders are simply eliminated altogether.

Despite some horror stories, E-mail offers a twofold advantage over alternative media such as voice mail, answering devices, and written correspondence. Compared with voice media, it speeds communication, since we listen to people speak at only 150 words per minute but read 250 or more words per minute. E-mail simplifies how you handle and respond to written mail. Another benefit lies in the computer's ability to sort through the electronic messages and pull out only those that are most important to you based on who the sender is or key words used in the message. It's like having an automated secretary who sorts your mail 24 hours a day.

E-mail can reduce paperwork in a myriad of ways. Consider the case of a major corporation that needs to communicate new prices and product information to its sales force. With E-mail, it can send the information to all salespeople nearly instantaneously so they are able to present up-to-the-minute information to their customers. That's how a major company in the insurance industry solved its problem of communicating with its large, independent sales force. The company, a major provider of health insurance and financial services, linked its internal system (PROFS) to a public-access system, MCI Mail. Agents were then encouraged to sign up for interactive MCI Mail accounts so they could receive product updates and administrative messages instantly and electronically.

Companies are cutting paperwork directly by automating previously paper-based forms with E-mail systems. That's what Hughes Aircraft is doing. It has set up an electronic forms routing and approval system. Expense account forms and hiring forms, for example, are sent through the E-mail network. These forms no longer accumulate or hide in someone's in-basket; managers approve or reject the electronic forms online. The system tracks the progress of each form through the system and records its ultimate resolution. The Hughes E-mail network, says one manager, "will not only lessen the glut of paper flowing through our mail rooms but also eliminate the need to retype forms as they move from one database or process to another."

You don't need an entire E-mail system to take advantage of electronic forms (sometimes called "E-forms"). The resource directory includes forms processing software, available at a reasonable price, that lets you automate business forms.

Electronic Messaging for Customer Support

One of the most innovative uses of E-mail is the notion of an electronic forum for customer service and support. Known as a "bulletin board service" or BBS, electronic forums allow customers to tap into online question-and-answer sessions 24 hours a day. A customer can type a question and later receive answers from the company rep as well as from other customers. In addition, a customer can read questions posed by other customers along with the answers, often obviating the need to ask their own question. "It's absolutely the best way to give support," claims Borland, considered an industry leader in online support.

It alleviates the problem of information overload in several ways. "It's a real important way to deal with the information overflow," one Borland manager explained to me, "both from the standpoint of the service provider like ourselves and the customer. When we post a message to *one* person, we're talking directly to that person while thousands of other people can see the answer. It's a one-to-many relationship.

"From the customer's perspective, when they come in and ask us a question, all these other customers are reading it and will respond to it too. The knowledge of the customer base is really huge. For example, a customer may call up with a very specific question about developing a Windows application. There may be three or four customers out there who have done exactly the same thing. They might tell them, 'Oh, I've done that before myself, and this is what I had a problem with. This is how I fixed it.' Or they might say, 'I've got some code you could have. Here, I'll upload it to the library, and you can download it.' Or they may even refer them to another forum from another vendor."

In terms of the financial impact of the forums, it's difficult to assess how the technology has affected the cost of providing phone support. The company has a volume of 60,000 to 80,000 calls per month. Perhaps it would be much higher without the electronic forums, but the impact is difficult, if not impossible, to assess.

Nevertheless, electronic forums clearly allow the company to leverage its efforts. Nearly 13,000 users from around the world tap into Borland's CompuServe forums *each month*. There's no way the company could duplicate internally the collective knowledge of that customer base.

Cutting Down on Computer Paper with EDI

Some estimate that nearly three fourths of the paper blizzard consists of computer printouts. Much of this output, in turn, becomes input to other computers. In fact, 70 percent of computer input is output from another computer. Generally it works like this. One company's computer outputs information on paper that is sent through the mail to another company, where the relevant information is input manually into that company's computer. For example, a supplier invoices its commercial customer, sending the invoice to both the customer and its own accounts receivable. The customer keys the information into their accounts payable system and, on the payment date, issues a check. The check and remittance advice are sent to the supplier, where an employee checks the information against unpaid items in accounts receivable and updates the account to paid status.

Electronic data interchange (EDI) eliminates the need for paper output and for rekeying the output as input to another computer. EDI electronically transfers the information directly from one computer to another and can be cost-effective for even small companies. Using EDI, for example, your customer could run its accounts payable system on Thursday. The funds automatically transfer to your bank account, *and* your accounts receivable system automatically updates that evening. During the night, the computer system routinely matches items to check whether the amounts paid match what was due. On Friday morning, your staff reviews any exceptions—cases where the amount paid did not match what was expected. This process not only saves time but is a more suitable use of human resources.

An estimated 10,000 companies are now using EDI, according to Eric Arnum, an industry analyst. Users now at the forefront of EDI include automotive, chemical, grocery, and pharmaceutical industries. Data "trading partners" can exchange directly, or a major bank or third party can serve as a buffer between the partners.

Implementing EDI is easier and less expensive than you might expect. The experience of Tennessee distributor Parrish-Keith-Simmons illustrates the point. With an up-front investment of only $3,000, the company implemented EDI by hooking into a third-party network (check the resource directory for EDI vendors) and began successfully exchanging data with a major customer within one week. "EDI has been a tremendous success for us," says company president W. Scott Parrish, "but in our wildest dreams we never anticipated how fantastic the results would be." To get started with EDI, check around to see if other departments in your company are already using it. Others to ask include the data processing group, internal or external auditors, professional groups such as the National Association of Accountants, or a major bank.

Filing "Electronic Paper"

Generally referred to as "image processing," the conversion of paper to electronic form cuts the time, space, and expense of storing and transmitting paper documents. With over 1 *billion* printouts, letters, and other office documents generated in the United States every day, the need for reducing some of that paper through image processing is obvious. Companies are responding to the need in record numbers—the image processing industry recently grew 68 percent and will continue to grow 50 percent each year for the foreseeable future. The trend to electronic paper will surely continue for many years, since most business data today is filed on paper, with only a small percentage filed *exclusively* on computer and the rest on microfiche or magnetic archives.

Business groups of all sizes can take advantage of the many ways to store paper in digital or optical form, because the costs to convert paper to electronic form have dropped substantially from previous levels. One company that switched to electronic paper is Motorola. Wanting to conserve trees, the company began putting its product catalog on computer disks last year. The electronic catalog lists over 13,000 products in 124 categories. Besides reducing the amount of paper, the electronic medium allows purchasers to search quickly and efficiently for the company's best product for a given application since the computer cross-references instantly.

New technologies make the transition from traditional files to electronic paper not only possible but cost effective.

Recordable CD and optical laserdisc. With a scanner and these new storage technologies, you can get rid of the paper pileups. One optical disc can hold 60,000 documents or the equivalent of three four-drawer traditional file cabinets. A major advantage of electronic paper is that it's much faster and easier to retrieve. That's why insurance companies and others who need quick access to large amounts of information are turning to electronic paper.

The property and casualty division of USAA now destroys 99 percent of all documents it receives, which amounts to enough paper to fill 39,000 square feet of filing cabinets every year. Instead of storing all that paper, they electronically scan all incoming documents—some 25 million a year—and store the resultant digitized pictures on optical discs the size of phonograph records. The computer system allows a policy service representative to instantly call up on his or her computer terminal electronic pictures of a customer's entire file. The company's CEO, Robert McDermott, says, "The system enriches the job by giving workers information right at their fingertips, so that they can make a decision on the first phone call." There's no need to run to the files to find an important letter or document. It's all "online."

Before the conversion it was a different story. "There was paper everywhere," McDermott said in an interview. "Every desk in the building was covered with stacks of paper—files, claim forms, applications, correspondence. You can't imagine how much paper. Stacks and piles and trays and baskets of it. And of course a lot of it got lost. On any given day, the chances were only 50–50 that we'd be able to put our hands on any particular file."

Microfilm and microfiche. These are not just for libraries and large archiving operations anymore. Termed "micrographics," this technology allows vast amounts of information to be stored in small areas at little cost. Micrographics can reduce the space needed for paper records—up to 98 percent—and allow fast retrieval. For smaller firms or individual departments, you can use a micrographics service bureau. These firms pick up documents or come in, film them, and create an index. You can find firms listed in your local business-to-business phone directory under "micrographics."

Bar codes are now being added to microfilmed documents to improve retrieval. Here's how it works. The bar codes are printed right on the documents to be filmed, or adhesive labels can be used. Then a special camera that encodes bar code information films the documents. The bar code number along with the microfilm access number are stored in a computer. A Computer Assisted Retrieval (CAR) system can be used to retrieve the needed documents more efficiently than can a regular microfilm reader.

Companies are saving big bucks by switching from paper to micrographics. Northrop, for example, has replaced paper with 36 microfiche cards in one of their aircraft assembly plants. They expect it to save some $20 million.

The biggest savings, however, may come in the form of space and peace of mind. One manager I interviewed told of how her staff complained about clutter. Order forms were stacked everywhere, and they were running out of space. As it turned out, the group had used a micrographics system in the past to archive paperwork but left the equipment behind when the company reorganized. The manager immediately ordered a new system to cut the clutter and restore peace of mind.

Files that Fit the Big Picture

Electronic media notwithstanding, paper will be with us for a long time. It's important to file information so you can find what you need later without getting overwhelmed with papers you don't need. As a manager, you probably delegate the task of filing to your staff or assistant. The best person, however, to make sure that files fit the big picture is you. The specific tools and techniques of filing are not as important as whether the file categories make sense for your activities and your business. As a manager, you are the best person to devise and periodically reassess your filing system categories and subcategories to make sure they are consistent with the big picture.

The subject of office files is important, because files contribute to the informational foundation of your business. Too often, this foundation has not been laid properly. One problem is that papers are saved needlessly, a common practice that wastes time, effort, space, and, ultimately, money. Another problem is misfiled papers that are costly to track down or reconstruct. Some lost files cannot be replaced at any price.

- By the year 2000, U.S. businesses will file 120 billion new sheets of paper a year, enough for 5 million filing cabinets.
- Nearly half of all file cabinets house duplicate records.
- The average firm could eliminate one third of their files altogether and transfer another third to inactive storage.
- Between 75 and 85 percent of filed papers are never referred to again.
- Lost files cost at least $60 per file to reconstruct.
- Each file cabinet you empty saves $600 a year.

The best way to streamline your office files, thereby saving time and money, is to keep them consistent with the big picture.

Early in my dual career, I learned it's costly to lose sight of the big picture. I had just completed an extensive market analysis for the marketing executive of a high-tech firm. I delivered the 135-page report by the project deadline and presented my recommendations with impressive graphics to senior management. But the effort left both offices buried under a disorganized mountain of paper, despite what I thought was an effective filing system.

It hurt to admit, but I needed help to get better organized. Determined to find out why my organizational scheme failed, I hired a firm to "overhaul" both offices. Thousands of dollars later, I discovered the problem.

The firm sent a specialist to my university office. Every piece of paper in that office related ultimately to one of three basic themes: teaching, R&D, or service activities. It was clear how all file categories fit into these areas. The specialist had no trouble putting everything in its place. For example, a set of slides belonged under Visual Aids, filed with Resources in the section on Teaching. The specialist easily finished the job in one day.

The consulting office, however, was a different story. I handed the specialist a list of some possible categories, but this time I could think of no basic themes around which to organize the categories. For the better part of three weeks, the specialist dutifully labeled files, rearranged information to prevent duplication, and threw away whatever seemed unimportant. In the process, however, valuable documents got lost or misfiled, and the consulting office quickly reverted to a disorganized state.

For months afterwards, I suffered from information overload. Ironically, I continued to conduct seminars and research on visual literacy and the importance of seeing the big picture. Then one day, the realization hit me that the problem with my consulting office was the lack of a guiding framework or perspective—there was no big picture. I had failed to practice what I preach.

Following my own advice, I developed a complete picture in my mind and on paper of the consulting business, the clients, the activities, and the organizational structure that resulted from this new perspective. Some of the material stored in my files turned out to be unrelated to this new structure and could be eliminated. When I brought the firm back in and supplied a framework to follow, the organizational effort was fruitful and lasting. The specialist could see where a particular item fit into the bigger scheme of things.

Devising a Framework

Invest some time to think about relevant categories that fit your big picture before you set up or revamp your filing system. You'll reap important dividends: save time looking for things, plan more effectively, and achieve control as your filing system becomes a vital database to back up your Master Control System.

"The best way to simplify complexity is through classification," advises Daniel Burrus, who researches global innovations in science and technology. He told me that he had spent a year coming up with a useful taxonomy of high technology when first starting his business. "I was fogged in with information. It was too complex. I couldn't make sense out of it." Old taxonomies weren't useful, because everything in the field of technology was changing so rapidly. He had little to go on since nobody had tried to organize new technologies. When I asked him if he developed a big-picture view first or developed the taxonomy first, he told me that the two evolved together. "Doing the taxonomy . . . helped to clarify the big picture."

So if you're not sure about your big picture, the task of developing a framework can help you clarify your thinking.

Filing Do's and Don'ts

Once you have devised the proper framework, you and your staff can take steps to keep files consistent with the big picture. Two helpful color booklets are *How to File Guide* from Esselte Pendaflex and *How to File and Find It* from the Quill Corporation (see the resource directory for ordering information). Share the following tips with your staff for better filing.

• Use hanging file folders because they allow modularization, save space, and make it easy to reorganize as your big picture changes.

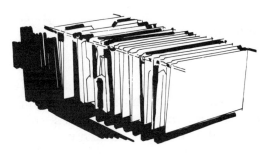

• Put metal drawer frames into the file drawers not already designed to hold hanging folders. Frames come with directions that show how to fit them into a drawer of any length.

• Label file folders with short, one or two-word titles. Categorical filing arranged alphabetically generally works best.

• Use tabs on hanging files. Tab the folder on the *front* flap. Arrange main categories on the left side, subheads in the middle, and individual categories such as company, product names, and people on the right.

• Place an item in a "hold" file if you're not sure of the category it belongs in. Review this file weekly. Additional time is often the best test of seeing how an item fits into the big picture.

• Printouts and booklets can also be stored in hanging files next to related documents. You'll find several styles of specially made folders with flat bottoms, with or without side flaps, to hold thick materials. The point is to keep related materials in one place for faster retrieval.

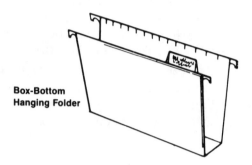

Box-Bottom Hanging Folder

• Store nonpaper information such as computer disks in specially designed hanging folders next to related paper files so that all information and resources connected with a project or category are in one place.

• Use 1/3 cut interior files for most purposes. This cut leaves enough room to label the file with names or geographic locations.

• Use 1/5 cut tabs for numerical filing. All numbers that are multiples of 5 go on the far right, and all numbers ending in 1 or 6 go on the left. That way all numbers ending in the same digit are always in the same position.

• You can switch to recycled filing products to decrease office paper waste and preserve resources. Recycled filing supplies include EarthSmart, EarthWise, EcoSafe, and other brands. See the resource directory for details.

Color-Coded Files

Color-coded files help you visually organize information. Color quickly guides you to specific categories, individuals, or activities. For example, use blue for current customers, red for high-priority files. According to Esselte Pendaflex, a major filing products manufacturer, color coding reduces filing time by 50 percent. (By the way, recycled products now come in an array of colors.)

• Pinpoint folders that need special attention. Key accounts, rush jobs, and confidential material in special colors such as bright yellow stand out instantly in a file drawer.

• Use color to identify different departments, numeric breakdowns, or time periods. Color helps you quickly spot different divisions and subjects in a file drawer, and you can go right to the part of the file you need.

• Spot misfiles quickly with color coding, because misfiling breaks the color pattern. An odd file or disk will stick out like a sore thumb.

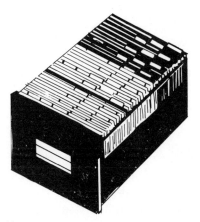

Misfiles are easily spotted with color coding.
Source: Quill Corp.

• Add a color dot to a file label to indicate a particular subgroup. For example, place a yellow dot on files that should be purged at the end of the year.

• Place color dots on audio and videocassettes (as well as other media) to tie media to specific projects or activities. All of the media associated with my basic seminar to accompany *SURVIVE Information Overload* have blue dots. Media for another seminar are flagged by a red dot. Materials used in both seminars have both dots.

• Draw a diagonal color mark across the top of a group of media such as a tray of 35mm slides or a set of computer disks (3.5 inch disks standing upright). You can tell at a glance if a disk or slide is missing from the group, or you can reorganize ones that have gotten mixed up.

• Use different colors of disks to flag different groups—blue for your word processing disks, red for the database set, and so on.

Color coding will help you track other files as well, such as card files. Harvey Mackay, author of *Swim with the Sharks without Being Eaten Alive*, uses color to separate business cards in his Rolodex file by function. "Customers are alphabetized behind the red divider, prospects are alphabetized behind the green, personal friends and relatives are blue, etc." He also uses different color ink each year to make notes so he can tell at a glance if he has talked to a person recently.

If you need help digging out from under and getting your files set up correctly, contact a professional organizer or call the National Association of Professional Organizers for a referral. See the resource directory for details.

Little Things Add Up

So-called little things can add up to big paperwork headaches. Chances are that a significant portion of your paperwork clutter is due to little things such as phone messages, business cards, notes you've written to yourself, receipts, and the like. These odds and ends contribute to the confusion, because they fall outside of the big-picture—they don't fit in. The solution is to integrate or connect them to some larger structure.

Eliminate all those phone message slips. Incoming phone calls can generate a steady stream of little slips that flood your office unless you take preventative measures.

• Avoid message buildup by scheduling incoming calls. Have your secretary ask callers to call back at specific times when you will be available to take the call.

• Delegate authority to your secretary to handle routine calls or page you for important calls. If you are not in, messages of important calls should be attached to relevant documentation before being given to you.

- Have your secretary log incoming calls on a special form, software, or spiral-bound message booklet. You can review the summary of calls quickly and decide which ones need a response from you or which can be delegated. It also provides a permanent record of incoming calls.
- Important calls can be logged on forms compatible with your Master Control System. Most suppliers of time planners and organizers provide phone forms.
- Use a stick-on note if something must be written on a small slip of paper. Then integrate it into the appropriate section of your MCS. Be careful, however, not to overdo it and jeopardize the streamlining benefits of the MCS.
- Add a voice mail or messaging system if you are not already using this technology. Listen to messages with your MCS in hand, so you can enter notes in the appropriate section. If it's important to get back to John tomorrow regarding project 5 in your MCS, write "Call: John—P5" on tomorrow's page of the daily calendar along with the phone number if it's not listed in the project or phone section. With one action, you've noted an important call, planned your future time, and made reference to supporting information.
- To prevent message buildup in your voice mailbox, ask callers to leave a succinct message. Or you could limit the time they have to speak and let them know so that nobody is cut off in midsentence. The outgoing announcement of one busy corporate professional I've worked with asks callers to leave a "brief but *specific* message." That way she can often handle the matter without first returning the call.

One of the best ways to handle incoming calls is to do a better job of managing outgoing calls. You'll learn more about that in Chapter 5.

Integrate stray business cards quickly. In addition to phone messages, business cards account for much of the clutter resulting from little things. Cards from business contacts collect on or in your desk, shelves, car dashboard, briefcase, drawers, billfold or purse, and many places other than the rotary card file where they belong. Stray business cards have plagued my offices.

As a seminar leader, consultant, and public speaker, I frequently receive business cards along with a request to send an article or booklet. Too many of those cards fell through the cracks before I adopted the following system for integrating business cards into the big picture.

- Write the date and occasion on the back of the person's business card as he or she hands it to you. Or if you are tied up with something, ask the person to write this information for you. (By the way, it is not appropriate to mark up a person's business card in some cultures so apply this tip accordingly.)

- Affix a special self-adhesive strip to the bottom of each card that makes it fit instantly into a rotary card file. During a business trip, you can do this on the plane. See the resource directory for product information.

- For key contacts, enter the information into a computerized database of names. A database allows you to manage your contacts more effectively. Lou Filippo, president of Sales and Management Development Company told me during a phone conversation that he could not run his sales and management consulting firm successfully without this electronic resource. "If you're not organized in this business, you are in trouble. I have a database that contains anybody I ever met. People I want to continue to have an association with in some form I put in my database."

"I'm looking at your file right now," he told me. "It has your name and address and brings up the last time I talked to you, the last time I tried to talk to you, and what your status or category is. I have you as a supporter rather than as a client or prospect."

For business travel, he selectively pulls out and prints four or five pages of relevant phone numbers so he doesn't have to carry an address book. Another frequent business traveler told me that he carries the information in his notebook computer. Then when he receives relevant business cards, he can simply enter them into the system on the plane trip home.

Several popular software packages for contact management are listed in the resource directory. See Chapter 5 for more information on this technology.

Box 3–1

Paperwork on the Road: How CEOs See It

• "I get the flight attendant to bring me a trash bag. I put it on the seat next to me and process paperwork. A long flight is often a two-bag flight."

Robert Crandall, *CEO, American Airlines*

• Tried a palmtop electronic organizer while traveling but spent too much time uploading and downloading. Switched to a notebook computer running PIM software, because it "can be set up the way you think."

Dan Hogan, *Principal, The Apollo Group*

• During overseas travel, phone calls back to the main office are prescheduled every few days. Anything urgent arrives by fax. Returns Friday to several packages of mail. "I'll be expected to go through them by the time I return to work Monday. What a weekend."

Daniel Gill, *CEO, Bausch & Lomb*

• Rather than travel to 21 offices in nine states, she has sales force fax daily report. "Simplicity is key. One piece of paper tells it all."

Marilyn Ounjian, *CEO, Careers USA*

ACTION PLAN: Take Back Your Desk Top Today

You won't see how items fit into the big picture when they are hidden somewhere in the stacks on your desk. The average executive wastes 45 minutes a day searching for something lost on a desk. If you spend just 30 minutes a day looking for information lost in the piles on your desk, you will have wasted more than *15 days* by the end of a year! The reward for taming your desk is a three-week vacation with no loss in productivity.

Make it your top priority to master the art of deskmanship. This does not mean simply tidying up your work space, as one manager discovered who directed employees to clear their desks at the end of each day. They began stuffing everything into drawers, leaving their desk tops clear. But the next morning, it took nearly an hour to find the information they needed to resume work, resulting in a predictable drop in productivity.

The following three steps help you clear your desk, see your way through the in-basket, and send paperwork along the pipeline. The common thread is to maintain a big-picture perspective on paperwork.

Step 1: C•L•E•A•R *Your Desk*

After you follow this five-point method to clear the clutter from your desk top, it's easier to achieve big-picture thinking because there are no visual distractions. Before I went through this process myself, I had difficulty spending more than a couple minutes reading reports at my desk, because phone messages, notes, and other papers sitting there—although less important—urgently cried out for a response.

Turn your desk from a distraction into a work surface by following this five-step process to CLEAR your desk top.

Clean it.   Clean off your entire desk surface by placing all stacks of paper and other items in a large box or on the floor. Replace only those items that you really do need on your desk top such as writing instruments, reference materials, phone, clock, etc. Once you've made a clean sweep of the clutter, leave enough open space in front of you to work on the task at hand.

If there are some papers that you simply must have on your desk top, organize them into files and store them in a vertical file box, which takes up little room but allows you to keep an eye on key papers. I generally keep summary papers for priority projects in a desk top file, because I refer to them frequently and don't want to dig them out of heavy desk drawers and filing cabinets each time.

Leave it for someone else to do.   As you go through your mound of clutter now on the floor or in a box, separate it into four piles corresponding to things you can delegate, eliminate, act on, or read.

Delegate as much as you can to your secretary or staff, especially if it relates to the organization's big picture but not your own. Anything outside of your Quadrants I and II is a candidate for delegation. When possible, refer paperwork to a colleague with greater knowledge or expertise in that area. For example, general requests can be routed to Public Relations, sales literature to Purchasing, and a legal matter to the Legal Department. Most filing can be left for someone else to do once the proper organizational structure is in place.

Eliminate it altogether.   Cut through clutter by asking, "How does this fit into the big picture?" Then throw away anything trivial or outdated. Start by eliminating items from Quadrants III and IV with no adverse effect. You can obtain copies later if it turns out to be important.

Examples of typical items to eliminate: most advertising brochures, last week's newspapers, sales literature, old clippings from magazines and newspapers, dusty articles, and the myriad little slips of paper that float around such as old messages, phone numbers, and personal notes. If related to the big picture, this information can be entered once in your MCS or stored in related files.

Act on it.   As you sorted through the clutter, you collected all items requiring your personal attention into an "action" pile or basket. If you did a good job on tasks two and three (delegating and eliminating), this stack should be manageable. Each item in the stack should relate to a particular project or category in your MCS. If not, recheck the item's importance and relevance to your big picture. Eliminate trivia that slipped through the previous stage, or delegate tasks that are not crucial to your big picture. You may need to redefine a category or add a new category to your basic file structure and MCS to accommodate an important item that doesn't seem to fit.

Box 3–2

The Case of a Cleared Desk

Edwina McLaughlin's open-door policy left her little time between meetings to return calls and tackle paperwork, let alone get organized. So she routinely came in early and stayed late to catch up.

Before interviewing job applicants, McLaughlin, the personnel director at a New York law firm, would sweep the unruly pile of paper on her desk into two empty drawers. But this Band-Aid solution failed to solve the deeper problem: her workspace chaos was encumbering her already frenetic workday. "I couldn't find my ideas on my desk," she says. She also was afraid to file anything away for fear she would forget about it.

Source: © 1991 Michael Grand

After a desk makeover, McLaughlin learned strategies for making her desk a tool for success by using it and her time more effectively. She became aware of how distracting her desk top was. Papers, problems, and priorities were competing for attention. Toning down the "visual noise" increased her productivity.

Order was imposed on McLaughlin's desk—and her workday—

with a personalized filing system: red "hot files" on the desk top for day-to-day priorities, project files on a nearby shelf, and less-active files stored in a drawer. McLaughlin no longer needs stacks of folders on her desk to remind her to do something; a To Do file in the hot section keeps her on track. A clip for message slips allows her to return calls all at once rather than interrupt herself throughout the day.

To free up desk top space, she had the phone attached to the wall and two shelves installed to hold Post-it notes, tape, and other necessities. McLaughlin used to waste precious minutes each day hunting down her electronic passkey when she needed to leave her

Source: © 1991 Michael Grand

office. (On the morning of the makeover, she locked herself out three times.) Now this and other essentials, such as business cards, have their own slots in a top-drawer organizer.

A clear desk gives McLaughlin more control. "I wasn't hopelessly disorganized before the makeover," she says, "but I no longer feel the pressure of the clock as intensely." McLaughlin doesn't have to work long hours to play catch-up, as she used to, because the moments once spent "getting it together" are now pockets of peace and quiet.

If you don't have time to act on an item now, at least note it in the MCS for later reference. Start an "Activities" sheet for each project or section in your MCS. Then list the required action on the Activities sheet so you won't lose track of it.

When I first CLEARed my desk, I discovered a collection of phone messages, unanswered correspondence, and other paperwork for one of the seminars that I conduct across the country. I promptly created a separate section in my MCS for the seminar, listed what needed to be done on an Activities sheet, and stored all of the seminar materials in a related office file. When I later followed up and saw the tasks on the Activities list, I delegated the matter to the marketing director, who booked several seminars—business that would have been lost if I had remained overloaded.

Items that require more immediate attention can be divided into two temporary groups—To Do Now and To Do in 2 Wks. Both groups of items can be placed in the vertical, desktop file. One woman I worked with found a number of magazine subscription notices after CLEARing her desk. Since she couldn't consider the choices and respond right then, she filed them in the temporary "To Do" file.

One tip: Get rid of your "To Do" files as soon as possible. In other words, relate *everything* to a larger project or area. I no longer use general "To Do" files, but I do keep a folder marked "Urgent" for pressing items that must be turned around that day or the next. The folder is see-through plastic housed at the front of my desktop vertical file. That way there's no chance it will be "out of sight, out of mind."

Read it.   Although reading requires "action," you should set aside in a separate category anything that takes more than a couple minutes to read. Reports, articles, magazines, and other publications can go in a basket or on a shelf to be read.

If there are many publications, divide them into categories: high/low priority, mandatory/optional, or professional/personal. Eliminate as much as possible, especially low-priority or optional reading. Toss anything more than three months old unless you have a compelling reason to keep it. Refer to chapter seven for more specific tips on how to cut down on reading time.

After CLEARing the desk top, it's a good idea to CLEAR desk drawers and all other surfaces. Go through the papers that have accumulated on bulletin boards, shelves, walls, and the sides of office furniture. One exception is inspirational material. I display quotes on my desk and nearby surfaces. My favorite: "People forget how fast you did a job, but they remember how well you did it." When my roving eye finds an inspirational quote, I'm renewed and return to the task at hand with added vigor.

Step 2: S•E•E through the In-Basket

As if one overflowing in-basket were not enough, I shuttled unopened mail from my in-basket at the university to one at my consulting office and back again until I developed this plan. I often found myself reading notices of upcoming events after they had occurred and opening letters after the requested response date. Now nothing hides in my in-baskets for more than one day. Here is the three-step plan to SEE your way out of in-basket pileup.

Screen then sort incoming paperwork. Before you receive mail or other paperwork, have your secretary or staff screen it for you. The goal is to whittle down your in-basket to important material only. If there's no available staff member to do the job, ask the mailroom clerk to remove junk mail before it's delivered to you (after you have made an effort to remove your name from such lists).

Next, sort incoming mail and other paperwork into categories. The goal is to integrate each item into your organizational framework as soon as possible. If you have an executive secretary, much of this task can be delegated.

Reports that relate to a particular project you're working on can go straight into the project file. Make a note on the project task sheet in your MCS to review it. Travel documents, brochures, and other correspondence about an upcoming business trip can be placed directly in the file for that event. Then you can set aside time to plan the trip and review all accumulated information.

Eliminate irrelevant papers. Look back at the Activity Analysis you conducted in Step 1 of the previous chapter. Papers that relate to Quadrants III and IV should be eliminated. Unfortunately, the discarded paper adds to our environmental problems. A more ecology-conscious method is to send a form postcard to each junk mailer requesting that your name be removed from their mailing list.

A more efficient approach to reducing junk mail is to contact a service that will remove your name for you. The Direct Marketing Association, for example, has a free service that will substantially reduce the amount of advertising mail you receive. More than 1 million people have signed up for the service. The 3,500-member group consists of companies anxious to know who throws their material in the trash so they can redirect their costly advertising effort to a more productive end. Send your name and complete address to: DMA Mail Preference Service, 11 West 42nd Street, PO Box 3861, New York, NY 10163-3861. Equifax Inc. of Atlanta offers a nationwide, $10 pick-and-choose service at 1-800-289-7658 that allows you to screen out certain categories of direct mail. See the resource directory for further information whether you are a consumer or a direct mailer.

Empty your in-basket daily.   It's impossible to respond to all incoming work the same day it arrives, but you can process it daily unless you're out of the office. (See "Paperwork on the Road" for tips to handle paperwork while traveling.) The goal is to integrate the contents of your in-basket into your organizational

"Sorry, but I don't have an "In" basket. I'm a recovering workaholic."
Source: © 1991 by Bradford Veley.

structure on a daily basis. That way, you can maintain a big-picture perspective over everything in your office.

If you've screened your mail and eliminated everything that falls in Quadrants III and IV, you should be able to process the contents of your in-basket in 10–15 minutes a day or less. The top priority is to integrate the contents of your in-basket into the big picture. Don't put it off for even a day or two, because it can pile up quickly and engender a vicious cycle: there's too much to handle, because I'm too busy, because there's too much to handle. . . .

Step 3: S•E•N•D It through the Pipeline

Once your desk is uncluttered and your in-basket is empty, you can see what's there to act on and SEND along. Here's a four-part plan for SENDing paperwork through the pipeline.

Set a deadline for all paper.   Why spend time reading a flier about a September seminar, when you wind up reading it in the middle of October? Reading last week's newspapers is a waste of time. The solution is to set a destruction date mentally or on paper. This technique applies to computer files and stored documents as well. Then make it a habit to glance at the date before you read anything. Is that news magazine you just picked up three months old? Pitch it and reach for the current issue instead. Is an old financial report sitting on your desk because you didn't have the time to go through all the numbers? File it if you need to refer back to the specifics later and spend your time on the current report.

Excerpt information.   You may not need an entire report, just one part. Ask the writer to send only the part you need, or remove the section you need and toss or return the rest. Request that anyone sending you a report write a brief executive summary of the main points. Then if you need more details, you can always request them later.

Another way to excerpt just what you need is to request "exception" reporting. What's the point in reading the same reports over and over every week or every month? Ask the author or producer of the report to bring only "exceptions" to your attention. You want to be informed if figures drop below a certain level or when new, competitive trends arise.

Network with others. Take care of paperwork without doing it yourself by forwarding it to someone more interested or better able to respond to the information. Then follow up to see if they've acted on it.

I make a habit of forwarding articles that I think will be of interest to my colleagues and contacts. It not only clears my desk but wins their appreciation that I was thinking of them and their concerns. For consulting clients, I often pass along noteworthy clippings when I send project reports and correspondence. A colleague of mine makes a point to send me clippings I can use in speeches. Sometimes it's a quote or interesting item like the ad I received last week for colored paper that claimed, "Color has power. Research suggests weightlifters actually lift less in pink rooms."

Weekly sales figures, technical reports, equipment records, and the like can be handled by others. Examine the paperwork you now receive and decide what you can redirect to your staff or colleagues. Then when you need specific data or information, you can tap the knowledge and expertise available in your network.

Decrease paperwork in stages. Transition to less paperwork by lengthening reporting intervals. If a report comes out weekly, ask for it to come out every other week. If you generate a monthly report, convert it to a quarterly basis.

Decrease paperwork by asking, "What is the worst thing that could happen if I stopped receiving this paperwork?" Survey results show that 43 percent of respondents said they were required to write unnecessary reports that contained information already available in other formats. Such reports were cited as the number one culprit behind unnecessary paperwork. Of course, make sure that a report is truly unnecessary before eliminating it altogether.

SELF-TEST: Are You Winning the Paperwork War?

                                                        Yes   *No*

1. Does it generally take you 30 seconds or less to find the name and phone number

of someone you need to contact in
another office?   ____ ____

2. Can you find any document on your desk
   top in a matter of seconds?   ____ ____

3. Can your assistant find any document in
   the office files within five minutes?   ____ ____

4. Do you keep only the latest issue of a
   magazine, journal, or newspaper near
   your desk?   ____ ____

5. Do you receive reports that you must
   read and reread in order to get the key
   points?   ____ ____

6. Do you bring paperwork home from the
   office more than once a week?   ____ ____

7. In the last week, did you spend more
   than 10 minutes trying to find something
   in your office?   ____ ____

8. In the last week, were there any papers
   on your desk, other than reference
   materials, that you did not refer to?   ____ ____

9. In the last three months, did you fail to
   answer an important letter because you
   misplaced it?   ____ ____

10. Are there more than two or three loose
    items such as business cards, phone
    messages, or notes on your desk top right
    now?   ____ ____

Scoring

Questions 1–4: 1 point for each "no."
Questions 5–10: 1 point for each "yes."

What Does Your Score Mean?

Score of 0–3   You are handling paperwork well. Skim this
               chapter for any new ideas such as converting
               to electronic paper.

Score 4–6   Your current paperwork load is causing some difficulty. After studying this chapter, invest the time to follow the Action Plan. When you reach the last chapter of this book, be sure to study the tips carefully and practice the five-step techniques.

Score 7–10   You feel particularly swamped with paperwork. Read and study this chapter thoroughly, making sure to follow the Action Plan. Before reading the rest of this book, skip to Chapter 7 on effective reading and writing. Study that chapter and spend the time to master the techniques offered in the Action Plan.

Coming up.   One reason people end up buried in paper-work is that in just about every field today, there is simply "too much to know." Find out how to "view the big picture" to stay ahead of the information explosion by turning to the next chapter.

Chapter Four

Too Much to Know? . . .
*View the Big Picture
(with Macros, Online Aids, and
Broadband Thinking)*

A moment's insight is sometimes worth a life's experience.

Holmes

You cannot send an ocean through pipes developed for a stream.

George Gilder

Decision makers today find an ocean of information when they look for a particular drop of data. The key to satisfying your thirst without drowning is simply this: *fiber-opticize to survive.*

The Fiber Optics Metaphor

Taken literally, fiber-opticize means that communication systems should be rebuilt with fiber optics to meet the needs of the information age. It would take 2,000 years to transmit the entire collection of the Library of Congress over our present telephone lines, but a fiber optic cable could accomplish the task in just eight hours! It doesn't matter if we find ways to push information through traditional wires faster. They are simply inferior to fiber optics and cannot keep up with our growing information needs.

Like packing data through copper wires faster, white-collar workers cannot keep up with the information explosion simply by working faster or longer. We need to interact with information in fundamentally new ways that are up to 100,000 times more effective! Traditional ways to deal with information—reading, listening, writing, talking—are painfully slow in comparison to "viewing the big picture."

- The average person speaks about 150–180 words a minute, writes in longhand about 20–30 words a minute, and types 40–70 words per minute.
- You read at the rate of about 250 words a minute if you are an average person reading average writing.
- Physiologists estimate the total capacity of the visual channel at 30–40 million bits of information per second. Assuming an average of eight bits per character and five characters per word, the capacity of the visual channel is equivalent to 45–60 million words per minute.
- For purposes of comparison, we'll assume that the readers of this book read 1,000 words a minute, well above the national average.
- Your ability, therefore, to view the big picture is 60,000 times faster than your ability to process information in traditional ways such as standard reading methods.

You can view the big picture through broadband thinking, which requires you to step back from the details and look for an overall pattern—the essence. Your status as an expert, specialist, or analyst—while desirable in the industrial era—can be your downfall in the information age. The reason is that the more steeped you are in a particular discipline, the harder it is to gain perspective on the biases and limitations the discipline affords. It also becomes more important to seek information from outside your area to solve particular problems within it.

Those who survive information overload will be those who search for information with broadband thinking but apply it with a single-minded focus.

Each of us is endowed with the capacity to focus on specifics as well as the capacity to view the big picture. We have the needed "wiring" to do both. There's no installation fee. In contrast, the costs and legal impediments to install an actual fiber optic system are substantial. The price tag to rewire the United States alone is at least $200 billion, and some put the figure as high as $1 trillion. Luckily, the only cost to fiber-opticize your thinking is the time you'll invest in this chapter and the rest of the book.

Too Much for One Person to Know

Most of us already face too much to know, because there is so much information on virtually any topic. The busy executive of a public utility complains, "I have to deal with 10 times more information than a decade ago." For this executive and many others like him, the situation is getting worse:

- Information is doubling every five years. By the end of the century, the amount of available information will double every 20 months.
- 50,000 books and 11,000 periodicals are published each year in the United States alone.
- More new information has been produced in the last 30 years than in the previous 5,000.

Nearly everyone is affected by this upsurge, including politicians. One member of the U.S. Congress admitted, "We're at a point where we're less informed about more and more that comes before us." Think about what the information explosion means to you. In five years, there will be five times as much to know in your area as there is today. If decision making is difficult now, how will you make good decisions in coming years?

Sharing Information

One of the fastest ways to see the big picture is to share information across departmental boundaries. Much of what you need to know to make effective decisions in your particular area generally comes from outside that department or function.

"Consider a business process such as order processing," says James Wetherbe, an expert in management information systems. "If the warehouse has five orders but only enough inventory to fill three, it must make a resource allocation decision. Typically, this decision would be made on a first-in/first-out (FIFO) basis. That seems equitable and fair, given the information they have available to them.

"This could result in a terrible decision. What if a customer who does a lot of business with the company really needs this shipment promptly, recently received an order late and was furious about it, is paying a high profit margin on the order, pays bills promptly, and a truck is routed to deliver a shipment to another customer nearby the same afternoon? But because a FIFO decision was made, the inventory is allocated to someone who hardly ever does business with the company, to whom the order is not urgent, who yields a low profit margin, does not pay bills on time, and a truck is not going into the vicinity for the next three weeks, during which time inventory could have been restocked anyway.

"Note that the information needed to improve the decision making in the warehouse comes from outside the warehouse. For example, customer need, importance, and profitability would come from sales, credit worthiness would come from credit, and shipping schedule would come from shipping.

"An organization that does not share information cross-functionally," concludes Wetherbe, "ends up with the left hand not knowing what the right hand is doing."

Everyone Is in Customer Service

The number one barrier to delivering excellent service, say managers and front-line employees, is inadequate communication between departments, according to a survey by Performance Research Associates. Departments can share information by simply picking up the phone and communicating. In large organizations, a computerized information system may be the answer. Information from each unit is integrated into the system. Then before each group makes a decision, they can check key information from other areas. By seeing the big picture, each group makes better decisions.

Giving every worker access to the big picture results in better handling of customer inquiries. Regardless of who the customer calls, that customer's questions can be answered. Responding to the customer on the first call makes sense from a company's big-picture point of view. But unless that view is fostered within functional units, each department will worry only about minimizing the time and effort its unit must expend as if it were disconnected from the whole. They will probably try to pass off customers to another department. That is why, for better customer service, it is essential to foster a view of the big picture.

Good heavens, I can't help you! I'm just a lowly paper pusher. You need to talk to somebody upstairs in our Paper-Shuffling Department.
© 1991 Brad Veley.

Teamwork and Collaboration

In the spirit of "two heads are better than one," teamwork and collaboration are essential in the information age. In companies across the United States and around the world, people are discovering the power of collaboration. "Our researchers are not any smarter," admits Mac Booth, CEO of Polaroid, "but by working together they get the value of each other's intelligence almost instantaneously."

Professional and technical workers are especially hard hit by the information explosion. A consortium studying the problem of information needs in science concluded that future success depends on "greater interdependence among all concerned—research libraries, the science disciplines, government agencies, commercial information providers, and scientists themselves."

One scientist, the director of veterinary services at the San Diego Zoo, laments his load. "I have to keep up with medical research on 900 different animal species." To stay current, he reads 32 monthly reports and 5 monthly journals. However, he simply cannot stay current in all areas. His solution is to hire outside consultants for those specialties.

Collaboration offers a solution, but it's not always easy to get technical workers to work together as a team. As a manager at 3M explained, "There is always a tendency for technical people to stay squirreled away in their own labs, concentrating only on their own division's technology." To encourage communication, 3M began a technical forum. "The effect of all these seminars and colloquia is to keep hundreds of fresh ideas bouncing back and forth among our laboratories all the time."

Another approach is to team technical workers with nontechnical people. Eastman Chemical Co., a division of Eastman Kodak, formed a "patent improvement team" made up of inventors, lab managers, and attorneys. As members of the team, lawyers discuss with scientists ways of tailoring research during the experimental phase to improve the likelihood that a patentable product or process will result. According to the company's vice president for development, teamwork has paid off: Patent submissions have increased by 60 percent, and the number of patents issued to the company each year has doubled.

Multidisciplinary Work Teams

Many of the largest, most successful companies are forming multi-department teams to foster both efficiency and effectiveness. By eliminating old hierarchical structures, team members can interact directly and, thereby, save time and boost creative synergy. Amoco Production Company, a subsidiary of Amoco Corporation, has begun to reorganize into multidisciplinary teams of 500 or so workers, shrinking the management hierarchy from six tiers to three. Each new group has authority to make decisions without going up the old hierarchy for approval. One team, for example, is made up of geologists, geophysicists, engineers, and computer scientists all working together to extract more oil from the Gulf of Mexico. The arrangement is working well by all measures. "We're finding more oil and getting better financial results with the same number of professionals and fewer managers," concludes the unit's manager.

Multidisciplinary teamwork helps speed products to market sooner. By creating cross-functional teams, GE's circuit breaker business overhauled its entire operation and slashed the order-to-finished-goods time from three weeks to three days while cutting costs 30 percent. "We'd be out of business if we hadn't done it," admitted a GE general manager. AT&T cut product development time for new phones from two years to one year by setting up multidisciplinary teams with authority to make every decision on how the product would work, look, be made, and cost. Honda, Motorola, Brunswick, Hewlett-Packard, and others have had similar success with cross-functional teamwork.

Teaming with Customers and Suppliers

Don't ignore customers and suppliers when looking for collaborators. In fact, if Apple can team with IBM, even competitors need not be overlooked.

Consider teaming up with your suppliers. That's what Chicago truck maker Navistar did. It had a major contract with U-Haul to deliver a new moving van fast. "We realized we couldn't deliver the truck fast enough in the traditional way," says the head of Navistar's truck building subsidiary. "It had to be done in teams, with everybody, suppliers included, working simultaneously." One of their suppliers helped design the new truck, and the approach succeeded. The company cut development time in half.

Many companies are beginning to see market research and customer service as something they can do *with* rather than *to* their customers. This is especially true in high-tech companies. Leading software companies turn to their best customers, their "beta sites," to find bugs and get ideas for improvement. Hardware vendors rely on their "user groups" for suggestions about future directions and hardware innovations. To team with the customer, technology and responsiveness are key.

• Networking with customers is still a novel idea but offers some of the same advantages of EDI with suppliers. A small mail-order company in Texas, for example, lets customers track the status of their orders by allowing them to tap into the company's production and shipping files remotely from the customer's computer. Another company, Global Village Communications, encourages new customers to self-register by using the newly purchased fax/modem hardware to call an 800 number that connects with the company's database. Enabling customers to electronically enter their own orders, provide data, or check the status of an order saves work for the company and gives the customer a sense of satisfaction.

• Electronic customer forums allow companies to tap the collective knowledge of the customer base. Currently used primarily by high-tech companies, electronic forums will offer the same advantages to any company in the future as E-mail and online usage continue to skyrocket in virtually every industry. Once a significant portion of its customer base goes online, a company can leverage its efforts with the one-to-many communication afforded by a forum. Customers become partners in responding to questions or problems aired by other customers.

• The goal is to move from summative evaluation, testing consumer reaction once a product is developed, to formative evaluation, getting feedback during development and modifying as you go. The latter type allows the customer to help with the formation of the product or service. Use a beta site, test bed, pilot testing, or include the customer on the development team.

In my role as professor, I supervise graduate students as they develop multimedia software and video projects. I require that the projects be put through a series of pilot tests and revisions. One student who created a video to train student nurses tested the reactions of several student nurses to successive versions of the video and revised it extensively based on their candid feedback. Another student who created a multimedia product to be used in Third World settings pilot-tested early versions of the product with foreign students from Third World nations, since they were most similar to the target market for his product. He made hundreds of specific revisions based on their feedback.

Hewlett-Packard's Multidisciplinary Map

Multidisciplinary work teams are proving valuable in areas such as product development. Manufacturing companies today are shortening development cycles through collaborative work of cross-functional development teams. One problem, however, in managing multidisciplinary teams has been that it's one thing to encourage collaboration between different functional groups, but it's another to try to evaluate performance cross-functionally. The folks at Hewlett-Packard, however, came up with a way to visualize team progress holistically and allow evaluation of all elements of the product development cycle as well. Called the "Return Map" (since it shows how long before the product shows a return on investment), their solution is "so simple and elegant that it has become a staple of the company's product development cycle," according to one director. It allows people from different disciplines to assess the impact of their decisions on the project in its entirety.

"The map tracks—in dollars and months—R&D and manufacturing investment, sales, and profit. At the same time, it provides the context for new metrics: Break-Even-Time (BET), Time-to-Market (TM), Break-Even-after-Release (BEAR), and the Return Factor (RF). These four metrics become the focus of management reviews, functional performance discussions, learning, and most important, they are the basis for judging overall product success."

The company provided the following example of the estimated and final Return Maps for development of an ultrasound machine. The first map plots estimates, and the final version shows what actually happened. The investigation phase was planned to last 5 months and cost $500,000, Time-to-Market was estimated to be 9 months and cost $2.3 million, while anticipated Break-Even-Time was 18 months. Mature sales volume was expected to be 300 units a month, or $16 million, and mature profits were expected to be $2 million per month.

An important benefit of the map is that it provides a visual perspective on the project. If one group wants to change things in midstream, it can see the impact of the changes on the entire project. That's just what happened during development of the ultrasound machine. Two months into development, Hewlett-Packard labs had a breakthrough in ultrasound technology that would enable the machine to offer clearer images. However, the change would boost development costs 40 percent and add four months to the Time-to-Market estimate. According to the Return Map, the Return Factor would actually be reduced slightly.

What initially looked like an open-and-shut case to incorporate a significant benefit turned out not to be a money-making idea. In the end, however, the group decided to use the new technology anyway despite their realization that it would not result in more product revenue. The team's purpose was to garner a larger market share, a strategic decision made with open eyes and an awareness of how the change would affect the bottom line in the short run.

A major benefit of using the map approach is that various "what if" scenarios can be tested. What if predictions varied by 10 percent, over or under? What has to happen to bring the product out six months earlier than planned? Each time you change one part of the map, its effects on everything else are summarized visually.

Another source of valuable information is the difference between the projected and final maps. If predictions seriously miss the mark, either the forecasting process or the development process is out of whack. Hewlett-Packard discovered that company managers were much more effective at predicting total engineering months than total calendar months. The root of the problem turned out to be "a tendency on the company's part to try to do too many projects with the available engineers—resulting in understaffed projects. Once management focused on staffing projects adequately, the company experienced a significant reduction in this kind of forecast error."

HP's corporate engineering director summarizes the benefits of the map: "The Return Map provides a visible superordinate goal for all the functions of the team and, in graphically representing the common task, helps them collaborate. No graph can substitute for judgment and experience—yet there is no substitute either for basing judgment on an accurate picture of experience."

During Investigation, the Ultrasound Team Makes Early Estimates

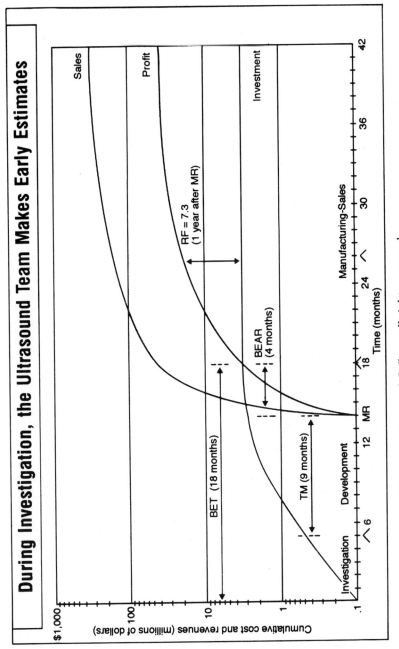

At Manufacturing Release, the Team Sees What Actually Happened

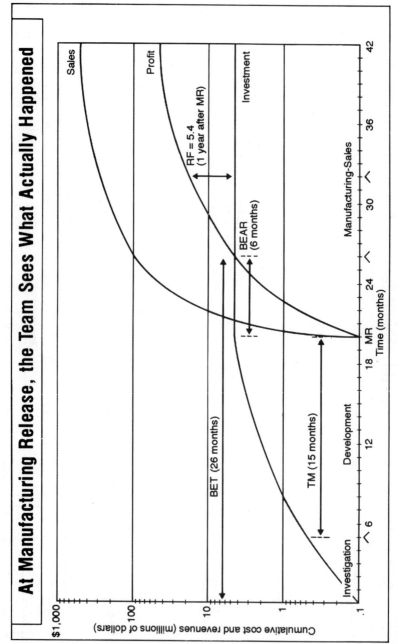

Visual Project Management

The Return Map was successful in part because it *visualized* the information in a way that helped everyone literally see the big picture. Team members could sketch the map on the back of envelopes during a discussion. Each person could mentally image how his or her efforts affected the overall effort. However, the Return Map is just one example of the many ways that visualization aids project management.

A graphic approach to project management is useful for virtually any project. The bigger the project, the more critical it becomes to use technology to visualize and computerize the project. "Even if the human mind was capable of comprehending the plethora of data involved in a complex project," explains Leslie Butterfield, vice president of Texim software, "in no way could one person visualize all the effects of a small change to a complex and dynamic plan.

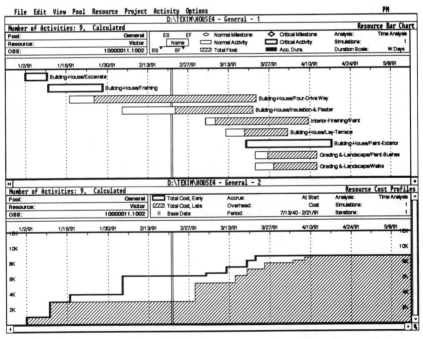

Graphical view of resource cost profiles and resource bar chart for complex project managed using *Texim Project*.

"Computers can recalculate data in seconds and provide a new view of the changed situation. The more views the software provides, the more likely the manager is to grasp all aspects of the plan, changed or unchanged.

"Software cannot take over the thinking, people management, or decision making necessary for running projects. But well-designed software can help in every stage of a project by analyzing available data and by providing a variety of ways of viewing the results to aid the project team in visualizing the plan."

Few companies, however, are using adequate project management tools, according to a survey of 500 executives and managers of technology-based companies by United Research, an international management consulting firm. Fewer than 20 percent of those surveyed were satisfied with project management efforts. More than half of the executives said that emphasis is placed on sequential events rather than on simultaneous development. Most of the survey respondents agreed that project management processes, tools, and training are sorely needed.

An easy way to manage your projects is to add project management forms to your Master Control System. Several suppliers offer specialized forms such as PERT/CPM sheets (Performance Evaluation and Review Technique/Critical Path Method). These forms present time requirements graphically so you can track simultaneous efforts. These forms keep you from getting overwhelmed in project details and help you see the big picture.

A Midwestern electronics firm realized dramatic benefits when it computerized project management to give everyone the big picture, according to *Planning Review*. The firm abandoned its sequential approach, in which each functional task was handled separately, and adopted an integrated approach. With the help of an outside consulting group, the firm broke the development project into 7,000 separate tasks that were tracked by an information system.

"It was an excellent communications tool because everyone could see when, where, and why the schedule was off track," reported the project coordinator. "If we saw that a unit or an individual was slipping behind, we could find out what was needed to get back on track." The end result was a 50 percent speedup in development time, lower development costs, and substantial improvements in product quality.

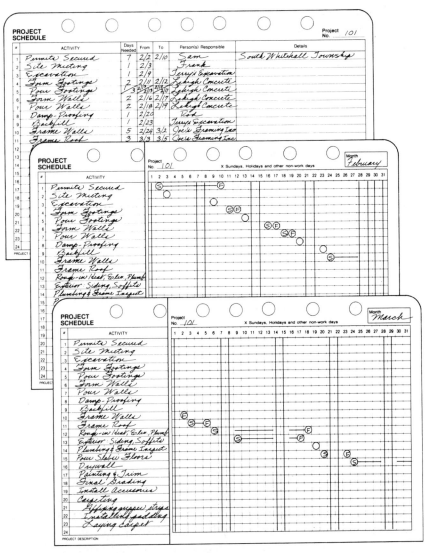

PERT/CPM sheets for project scheduling.
Source: Illustration courtesy of Day-Timers, Inc.

Why Is Visualization So Valuable?

In my work as a researcher, I have written several large, integrative reviews of the effects of all types of visualization—graphics, charts and diagrams, photographs, drawings, mental imagery, and other related approaches to visual communication. The evidence supporting visualization is overwhelming. The following five characteristics of visualization summarize why survival is spelled I-M-A-G-E.

1. "I" Is for Instant

Images have an *instant* impact on us. To illustrate this point in seminars, I flash a table of sales figures for three products for only a few seconds and then ask participants to identify the product characterized by a fluctuating sales cycle. Next, I quickly project a

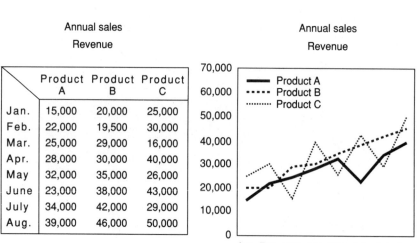

	Annual sales Revenue		
	Product A	Product B	Product C
Jan.	15,000	20,000	25,000
Feb.	22,000	19,500	30,000
Mar.	25,000	29,000	16,000
Apr.	28,000	30,000	40,000
May	32,000	35,000	26,000
June	23,000	38,000	43,000
July	34,000	42,000	29,000
Aug.	39,000	46,000	50,000

graph of the data. Generally, it takes much longer than a few seconds for someone to spot the trend from the numbers. But nearly everyone can instantly see the cycle when graphically displayed in color. The point is that images instantly communicate, while words and numbers are slower to release their messages.

The instant nature of images has major implications for surviving overload, as companies are discovering. A marketing manager in the automotive industry recounts how visualization simplified his job. In order to determine where to locate new dealerships, "a truck came with all this paper one day, and there it sat for a year." After the company decided to visualize the data, a map was plotted to reflect the many variables: location of existing dealerships, density of existing owners in the region, number of similar cars in the area, and so forth. The manager could quickly pinpoint a target location by simply looking to see where there was a large concentration of current and potential owners with no dealership to serve them.

Other companies have discovered similar benefits. Before visualization, the director of office systems at an insurance company sent out market analysis reports that were literally "three feet high" and that "went into a bookcase for a year until the new one came out." Now the same manager uses graphics to map the information. As he explains, "Boom. Put it on a map and you can see it in 15 seconds."

2. "M" Is for Memorable

My research and hundreds of other studies have shown that visual information is easier to remember than verbal information. Try the following activity to see for yourself if you remember images more easily than words.

Some years ago, a television commercial for a popular brand of margarine showed "Mother Nature" in a long, flowing white gown. As she warned, "It's not nice to fool Mother Nature," lightning punctuated the swirling storm. Can you picture the ad? Most people can. Here's the hard part: Can you name the *brand* of margarine it advertised?

Another margarine commercial showed a person taking a bite of bread, a coronet fanfare played, then a crown magically appeared on the person's head. What's the brand? Unlike the first ad, most people do remember—Imperial.

I've conducted this exercise with literally thousands of people in seminars across the country with the same results each time. A handful of people in each group invariably claim to know the first brand (then argue over whether it was Parkay, Fleischmann's, or Chiffon) while nearly everyone recalls Imperial. Why the difference? Both ads show clever, memorable images. But only the Imperial ad shows an image that is closely tied to the brand name.

I've used this activity with a variety of audiences—executives, doctors, college students—to illustrate that visual images are memorable, but words are soon forgotten unless they are linked to something visual.

3. "A" Is for Automatic

A person must exert effort to extract the meaning from words and numbers; verbal communications must be read and reread to be understood. With images, however, understanding occurs almost automatically simply by viewing. Advertisers know this and get around FTC regulations against deceptive advertising claims by "saying" it visually. The typical Marlboro ad is a case in point. "The copy, of course, says very little," according to the *Media Industry Newsletter*. "The visuals, exquisite, artful, eye-filling, say a great deal. We hardly bother to read it, because we are so affected by the background of sweeping plains, majestic hills, a chuck wagon and campfire at night. A dream of paradise."

Verbal information, whether written or spoken, takes effort to understand and is open to multiple interpretation. Images, in contrast, are automatically understood and help ensure that everyone gets the same message.

4. "G" Is for Global

As Hewlett-Packard found with their Return Map, images allow team members to share a common, global view of a project. The more complex a project, the more important it becomes to provide a global view for all involved. A Canadian software design team discovered the importance and global nature of images during a complex software design project. In past projects, the Canadian team treated diagrams as helpful supplements to written communications—pictures were pretty but not pivotal. The project director explains, "We realized that we could concisely communicate a tremendous amount of conceptual information by using diagrams illustrated by text, rather than by using text illustrated by diagrams." The difference between the two approaches may seem trivial, but it had a substantial impact.

"We were able to communicate very effectively with people about the system through the diagrams," according to the manager. Everyone associated with the project understood what was going on. "One of our technical writers returned a document he had edited, and remarked that he was getting really interested in the system; it was the first time he had actually fully understood the content of a document that he had edited since being with the company."

5. "E" Is for Energizing

Images impact the emotions and energize people to action in a way that words seldom can. The executive director of the General Mills Foundation, Dr. Reatha King, uses visualization during fundraising appeals. "Vision is very much involved in selling philanthropy," says King. "I describe the needs I have seen at the agencies I visit, then I paint a picture of how things could be different with people's help."

During a visit to a state child-care center, the director told King, "We have some terrible problems. You can see it from the way the children respond to me. If I put my finger in the hand of a healthy baby, he'll respond by grasping it. If I put my finger in the hand of a cocaine-addicted baby, there is no response. The baby has no grip." King used this story of babies who "don't have a firm grip on life" to urge potential donors to "come to grips" with the situation. The result of the talk was a significant grant.

Are You a Big-Picture Thinker?

Each of us has the capacity to focus on details, and each can see the gestalt or bigger picture. A metaphor for understanding these dual psychological capabilities has been the physiology of brain hemispheres, each specialized in different types of thinking. Brain researchers have concluded that the left half of the cerebral cortex, the left hemisphere, generally pays attention to details, while the right side sees the overall pattern. The earlier, popular view of the verbal left brain and the visual right brain has been called into question and labeled a gross oversimplification by more recent findings. Irrespective of physiology, two different ways of thinking have been documented:

Focused	Big-Picture
Analysis	Synthesis
Verbal	Visual
Details	Gestalt
Sequential	All-at-once
Parts	Whole

Despite the fact that everyone can think in these two different ways, individuals often rely on one style called their *cognitive style*. Lawyers, for example, differ markedly from artists on the basis of cognitive style. The most obvious difference between lawyers and artists, of course, is that lawyers communicate with words, while the artist's vocabulary consists of visual forms. In a study of cognitive style, sculptors scored better than any other group on a gestalt figure-completion test that requires recognition of partially obliterated silhouettes. The gestalt test forces the brain to synthesize in order to get the right answer—build up a picture in the mind's eye of the final picture by looking at specific pieces. Lawyers performed rather poorly on this test, although they did score higher than any other group on a test of verbal-analytic ability taken from a popular IQ battery.

Another study of cognitive style compared executives and operations research analysts during a verbal-analytic task and a visual-holistic task. Overall, both groups adapted their thinking patterns, at least to some extent, to the task at hand. Verbal-analytic tasks elicited left-brain activity, whereas the right brain became more active during visual-holistic tasks.

The groups differed, however, in the degree to which they were able to adapt. Analysts displayed less right-hemisphere thinking than did the executives during the visual-holistic task. Evidently, the analysts were so steeped in analytical thinking that they relied on left-brain processing even when it was not appropriate.

Experts in management decision making are calling for the balanced use of both sides of the brain. "Complex decisions require

Item from the Street Gestalt Figure-completion test. (Correct answer is "rabbit.")

richer information support that engages the right and left cerebral hemispheres. For many successful managers at top levels, this implies the balanced use of both hemispheres."

ACTION PLAN: Seeing the Big Picture

Step 1—Build a Macro Reference Shelf

Lighten your information load by keeping a shelf of key reference aids that provide macro views of business information. As Samuel Johnson said, "Knowledge is of two kinds. We know a subject ourselves or we know where to find information upon it." A well-stocked reference shelf helps you do the latter, especially if it aids global thinking. Here's a list of macro reference aids to get your started:

An association directory. One of the fastest ways to get needed information is to ask someone who already has a view of the big picture. A helpful resource is an association guide that lists the associations for every industry. There is literally an association or expert on virtually any topic. One call to these people can save you hours of time. Seena Sharp, an information search specialist, often relies on experts before turning to other sources. "I've spent probably 70 percent of my time on the phone," she said in an interview, "ferreting out the information, talking to editors, talking to all those different people in the trade."

I keep a copy of *The National Directory* on my reference shelf. It's compact—the size of an average phone book—unlike the voluminous *Directory of Associations*, meant for libraries. Yet it includes several hundred thousand listings of associations, corporations, and government organizations. Yesterday, I needed to check the latest statistics on magazine publishing in the United States, so I called the Magazine Publishers of America, which I found in the *Directory*.

A current almanac. According to *Mastering the Information Age*, "Almanacs are perhaps the best information resources for the money." A $10 bill buys you a lot of facts in almanacs like *Information Please* and *The Universal Almanac*. In the time it takes to look up "temperature" in the index and turn to page 403, you can find the average temperature in January for Tokyo or London. Calling your travel agent for the same information or looking it up electronically could take twice as long. Wondering about college tuition rates at different schools for your son or daughter? Flip to the section on colleges to see a complete overview of accredited institutions with tuition ranging from nearly $20,000 at some private schools to under $1,000 at the institution where I teach. You'll even find a one-page profile of the *world*—our GWP (gross world product) is $20.3 trillion, growing at 3 percent annually.

A company profiler. One of the most used references on my shelf is *Hoover's Handbook*. There are American Business and World Business editions. In one-page overviews, this handbook gives you a big-picture view of the world's top firms: the company's background, milestones in its history, who's who in the organization, its ranking in the United States and in the world, visualized financial statistics, and commentary with interpretation of what's been happening there lately. You could get this type of information electronically, but online services can be expensive compared to the wealth of information this book provides for

SONY CORPORATION

OVERVIEW

The name Sony is derived from *sonus*, the Latin word for sound, symbolic of the company's early successes and Chairman Akio Morita's affinity for music. Sony is a world leader in consumer electronics, video technology, recordings, and films, generating about 70% of its sales outside of Japan. It owns 52.5% of audio equipment manufacturer Aiwa.

Spending over $1 billion on R&D, Sony has traditionally sold innovative products at high prices rather than compete on price with commodity products. However, after consistently losing market leadership to price- and cost-cutting rivals, Sony has become more price

conscious. The company has also moved into less intensely competitive industrial markets.

Sony recently combined its entertainment businesses (including CBS Records and Columbia Pictures) into US-based Sony Software, which will develop and market entertainment software in formats suitable for Sony's new hardware technologies, such as high-definition television and 8mm video.

Chairman Morita tarnished his sterling US image by contributing to an America-bashing book in 1989. He has withdrawn his contribution from the 1991 English language version.

WHEN

Akio Morita, Masaru Ibuka, and Tamon Maeda, Ibuka's father-in-law, established Tokyo Telecommunications Engineering in 1946 with funding from Morita's father's sake business. Determined to innovate and create new markets, the company produced the first Japanese tape recorder (1950).

In 1953 Morita paid Western Electric (US) $25,000 for transistor technology licenses — a move that sparked a consumer electronics revolution in Japan. His company launched one of the world's first transistor radios in 1955, followed by the first Sony-trademarked product, a pocket-sized radio, in 1957. The company changed its name to Sony in 1958. It introduced the first transistor TV (1959) and the first solid-state video tape recorder (1961). Sony preempted the competition, becoming a leader in these newly emerging markets.

Morita moved to New York in 1960 to oversee US expansion. Sony launched the first home video recorder (1964), solid-state condenser microphone (1965), and integrated circuit–based radio (1966). Sony's 1968 introduction of the Trinitron color TV tube began another decade of explosive growth. Its Betamax VCR (1976) was defeated in the marketplace by products employing rival Matsushita's VHS technology. The Walkman (1979), in all its forms, was another Sony success.

By 1980 Sony faced an appreciating yen and intense price and quality competition, especially from developing Far Eastern countries. The company began to use its technology base to diversify outside consumer electronics, and it began to move production to other countries to reduce the effects of currency fluctuations. In the 1980s Sony introduced Japan's first 32-bit workstation and became a major producer of semiconductors, audio/video chips and components, and floppy disk drives. The company remained active in the consumer market, introducing the Watchman and an 8mm camcorder, developing CD technology with Philips, and becoming the market leader in CD players. Sony made large investments in digital audiotape (DAT) technology and high-definition television.

Sony acquired CBS Records from CBS for $2 billion in 1988 and Columbia Pictures from Coca-Cola in 1989 for $4.9 billion. The purchases made Sony a major force in the rapidly growing entertainment industry and gave the company control of a large library of films, television programs, and recordings. In 1991 the company signed superstar Michael Jackson to a new record and film contract (the biggest ever in the industry). Sony had its best year ever in fiscal 1991, with sales up nearly 26% and profits up nearly 14%.

HOW MUCH

Stock prices are for ADRs ADR = 1 share	9-Year Growth	1981	1982	1983	1984	1985	1986	1987	1988	1989	1990
Sales ($ mil.)	17.9%	4,266	4,863	4,578	4,804	5,211	6,788	8,309	11,655	16,678	18,760
Net income ($ mil.)	8.1%	325	307	186	127	292	344	259	294	549	655
Income as % of sales	—	7.6%	6.3%	4.1%	2.7%	5.6%	5.1%	3.1%	2.5%	3.3%	3.5%
Earnings per share ($)	2.9%	1.51	1.34	0.81	0.55	1.18	1.38	1.04	1.15	1.82	1.95
Stock price – high ($)	—	26.13	18.00	16.75	17.38	21.38	23.50	40.25	58.50	65.75	61.50
Stock price – low ($)	—	14.50	11.00	12.63	12.75	13.50	18.13	18.25	35.38	49.75	40.25
Stock price – close ($)	10.5%	17.50	15.25	15.63	14.00	20.38	20.50	37.75	57.88	60.50	43.00
P/E – high	—	17	13	21	32	18	17	39	51	36	32
P/E – low	—	10	8	16	23	11	13	18	31	27	21
Dividends per share ($)	10.2%	0.14	0.15	0.18	0.18	0.18	0.20	0.28	0.33	0.34	0.34
Book value per share ($)	16.1%	7.15	8.47	8.36	8.85	9.58	12.23	16.19	21.79	24.44	27.44

1990 Year-end:
Debt ratio: 31.1%
Return on equity: 7.5%
Cash (mil.): $4,388
Current ratio: 1.10
Long-term debt (mil.): $4,114
No. of shares (mil.): 332
Dividends:
1990 average yield: 0.8%
1990 payout: 17.4%
Market value (mil.): $14,273

Stock Price History
High/Low 1981–90

NYSE symbol: SNE (ADR)
Fiscal year ends: March 31
Hoover's Rating **A-**

WHO

Chairman: Akio Morita, age 70
President and CEO: Norio Ohga
Deputy President: Masaaki Morita
Deputy President: Nobuo Kanoi
Deputy President (Finance): Ken Iwaki
Deputy President (Personnel): Tsunao Hashimoto
Auditors: Price Waterhouse
Employees: 95,600

WHERE

HQ: 7-35, Kitashinagawa 6-chome, Shinagawa-ku, Tokyo 141, Japan
Phone: 011-81-3-3448-2111
Fax: 011-81-3-3448-2244
US HQ: Sony Corp. of America, Sony Dr., Park Ridge, NJ 07656
US Phone: 201-930-6440
US Fax: 201-358-4058

Sony operates worldwide.

	1990 Sales	
	$ mil.	% of total
Japan	5,538	30
US	5,464	30
Europe	4,557	25
Other countries	2,785	15
Adjustments	416	—
Total	**18,760**	**100**

WHAT

	1990 Sales	
	$ mil.	% of total
Video equipment & TVs	7,581	41
Audio equipment	4,600	25
Records	2,900	16
Movies	589	3
Other products	2,674	15
Adjustments	416	—
Total	**18,760**	**100**

Consumer Brands	
Betamax	TVs & VCRs
Data Discman	Videodisc players
Discman	Videotapes
Mini Disc	
Sony	**Commercial/Industrial Products**
Trinitron	Airborne audio/visual systems
Walkman	Commercial display systems
Watchman	Computer disks and drives
	Batteries
Consumer Products	Digital position readout sys.
Audio systems	Optical disks and laserdiscs
Audiotapes	PCs and workstations
Camcorders	Semiconductors
Car video systems	Sputtering equipment
CD players	
Prerecorded music	**Entertainment**
Still-image video	Sony Pictures Entertainment
cameras	(formerly Columbia Pictures)
	Sony Music (formerly CBS Records)

KEY COMPETITORS

BASF	3M	Rank
Bertelsmann	Motorola	Samsung
Canon	NEC	Sharp
Daewoo	Oki	Thomson SA
Fuji Photo	Olivetti	Thorn EMI
Fujitsu	Paramount	Time Warner
Hitachi	Philips	Toshiba
Lucky-Goldstar	Pioneer	Walt Disney
Matsushita	Polaroid	Zenith

One-page company profile from *Hoover's Handbook.*
Source: Illustration courtesy of the Reference Press.

under $25. I often spend a few minutes reviewing a company's profile right before any business contact to fix *their* big picture in my mind. They're generally impressed that I have a feel for their business beyond our specific business dealings.

A guidebook to finding information.   This type of resource is a must in the information age. There's *Find It Fast,* which covers both conventional information sources and electronic databases. *Lesko's Info-Power* is an award-winning reference book that I've used to locate experts. In one section, it lists names and phone numbers for thousands of experts organized by topic. I called the man listed as the expert on the Paperwork Reduction Act. Although he turned out not to be the exact person I needed to talk with, he knew all about the act and could refer me to just the right individual. Without his name and number, I surely would have made eight or ten calls to get to the right person.

An illustrated dictionary and other writing aids.   With the rising popularity of spell-checker software, traditional dictionaries are used to check meaning more often than spelling. Picture dictionaries are not just for kids anymore, because pictures help you learn, understand, and remember meanings. I keep an illustrated dictionary on my reference shelf next to an English grammar and composition guide. A thesaurus is useful, too, although many people prefer the computerized version because it's faster and allows successive searches.

The most unusual but effective dictionary I know for checking meanings is the *Reader's Digest Reverse Dictionary* or their new illustrated version. Instead of looking up a word to find its meaning, you look up the meaning to find its word. For example, if you can't remember what to call those dots (. . .) in punctuation, you would look up *dot-dot-dot* to find the term *ellipsis.* The term is also listed under *punctuation.* Look up *paragraph* and you would find the definition for *indent:* "Set a paragraph in from the margin." Want to know what the end section of a bound book is called? You'll find the term *headcap* under *bookbinding.*

Handbooks in your specialty.   In virtually every field there are handbooks that consolidate a vast amount of information pertinent to your area. I use *Writers Market, Speaker's & Toastmaster's*

Handbook, and the *Complete Speaker's Almanac,* among others. To maintain a global perspective, I look for resources that reflect cultural diversity. Check the reference section of the bookstore, or ask your professional or trade association to send their catalog.

Business and trade guides and directories. In addition to the local business phone book, my reference shelf holds the *AT&T Toll-Free 800 Directory,* current directories of professional and trade organizations, resource directories, and buyers' guides of products and services. Some of these numbers will be in your rotary card file, but it saves time to keep a directory handy for those that are not.

Reference materials on disk. The office of the future can be yours today. The extensive storage capacity of CD-ROM technology (similar to the audio CD) puts an enormous amount of information at your computer's fingertips. There are already many reference materials on CD and the list of titles is growing steadily. You'll find an encyclopedia, world atlas, almanac, thesaurus, magazines, and much more on CD. Unlike paper-based reference materials, these aids include sound effects, color graphics, and sometimes even animation and special effects.

Step 2—Get the Big Picture with Online Resources

Although traditional reference aids have the appeal of simplicity and affordability, nothing can beat electronic information (called "online" since you tie in to a mainframe computer) in scope and timeliness. Actually, most electronic information is published periodically in book form but in a size and bulk appropriate for libraries, not your desk. It works the other way too: just about everything in a library these days is stored electronically— somewhere. The challenge is to target the right information quickly since the fee to use online information is generally $1–2 *per minute.*

To use online services, you need a computer equipped with a modem (a device for connecting a computer to a phone line). Once you're hooked up, you can access a veritable cornucopia of information organized into collections called "databases." An electronic database is a collection of information related to a specific topic.

Electronic aids can save you from information overload but, ironically, could make matters worse by adding to the confusion. The first challenge in using electronic or online reference aids is to figure out which one to use. There are literally thousands to pick from. Alfred Glossbrenner says, "With so much information now online, it is exceptionally easy to simply dive in and drown."

Determine what type of database you need. A database generally consists of either "source" information (actual data or full text) or "bibliographic" information (source summaries or references). *Harvard Business Review,* for example, is a source database that contains the complete text of the journal's contents. *Arthur D. Little/Online* is a bibliographic database that includes a reference and summary of each article but does not include the article itself.

To find the information you need from a database, you must know which database you want to search, how to enter key words to retrieve information from the database, and how to save or store

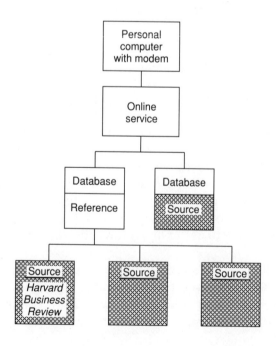

the information once it has been retrieved. Some services allow you to search many different databases at once for the information you want. There are several drawbacks and limitations to online information searching: *(a)* the cost of subscribing, searching for, and printing data; *(b)* the time it takes to learn how to use the service and/or particular database; and *(c)* the time it takes to sift through retrieved information to pinpoint what you need, especially if you have not targeted your information request. Of course, there are a variety of products and services available to minimize these limitations.

   Select a gateway service, then a database (unless you already know which database you want). Gateway services provide access to a variety of databases. Two of the most popular services are DIALOG and CompuServe, which are listed in the resource directory. They give you entree to hundreds of commercial databases as well as to bulletin board services (BBS), electronic catalogs, and much more.

   It is not mandatory to access databases through a gateway service. You could tap into a particular database directly and arrange payment with the database provider. In practice, however, a gateway service generally makes more sense. The reason is that once you begin using online services, it's likely that you'll want to use resources from many different databases. You can even earn a college degree online. With a gateway service, you have access to a variety of choices and are billed only by the gateway service. Here are just a few examples of databases that you can access online.

- *Standard and Poor's Register.* Their *Biographical* database includes personal and professional information about 70,000 executives in private and public companies. Their *Corporate Descriptions* database provides important business facts and profiles on leading public companies.
- The *Thomas Register Online* contains information on what is made in the United States, where it is made, and who makes it. The file covers over 146,000 U.S. manufacturers, with over 50,000 classes of products and over 110,000 trade or brand names.
- *Moody's Corporate Profiles* has descriptive profiles and financial overviews of over 5,000 publicly held companies. The information is based on news releases, articles in

newspapers and magazines, annual reports, SEC filings, and other public agency data.

- *Disclosure* provides in-depth financial information from SEC and other records on over 12,500 companies, including commentary on significant events and conditions affecting each company.

- *Company Intelligence* gives the standard information plus the latest 10 news references on over 100,000 public and private companies. If your company or a competitor has been in the news lately, this is where you can find out about it.

- *Findex* summarizes and references industry and market research reports, studies, and surveys that are commercially available from United States and international publishers. If you were planning to do any market research or conduct a survey, check here first to see if the information is already available.

There are databases for virtually every industry and profession. For lawyers, there's *Legal Resource Index, Laborlaw, EBIS—Employee Benefits InfoSource,* and *CDB Infotek's Investigative Information System,* among others. For doctors and health care professionals, there's *Medline,* which corresponds to three major print indexes in medicine and dentistry. In science and technology, you'll find *SciSearch, Compendex Plus, Inspec, NTIS, Energy Science and Technology, Aerospace Database,* and many others.

Invest in training and other online aids. Using a database is a matter of connecting with a mainframe computer and "telling" it what information to search for. The process is referred to as "conducting an online search." As you might guess, it's not that easy to communicate with a computer. Some databases make it easier because they are controlled by a "menu," an on-screen selection of possible choices. You conduct your search by picking various alternatives from the menu. Although menu-driven databases are easier to use, they are also more expensive than command-driven databases since menus generally take longer to use. Commands, on the other hand, take longer to learn initially.

There are a host of aids to help you use online services more efficiently. The major services have training manuals and resource guides. As one CompuServe user put it, "Get a few CompuServe-related books including the *CompuServe Almanac.* The books are not an option but a necessity."

With thousands of online resources to pick from, you face the bewildering task of selecting the right ones. To help you figure out the right databases for you, the major services also offer seminars tailored to business users and other specialists (medical, legal, etc.). First you learn how to use the general service, then how to use each database you're interested in.

Software products and services are available to help you search more efficiently. With a search-aid program, you plan the search on your personal computer before you access a database and start to accrue online charges. *ProSearch,* for example, is a search-aid program for retrieving information from DIALOG databases. It's adaptable to both novices and experienced searchers. It has saved me a bundle in online fees. I use it together with their software that automatically formats the information once it is retrieved.

You can usually "download" (transfer software from the online service to your personal computer) programs that help you plan more efficient online sessions before going online. For example, IBM users accessing databases through CompuServe can copy a program called ATOSIG. Using this software, you plan your session while offline. You could tell the software to check news in particular forums (electronic bulletin boards) or retrieve your E-mail. The program then automatically goes online and efficiently retrieves the desired information. To save online charges, it disconnects as soon as everything has been retrieved, and you can read the retrieved information at your leisure without incurring additional charges.

Some Dos and Don'ts of going online. *Do participate in online forums.* There are thousands of forums or roundtables for users to exchange E-mail, participate in the online equivalent of conference calls, and share computer files. You'll find forums for major software programs, hardware, business topics, hobbies, and virtually any other area of interest. The president of a software company advises new CompuServe users: "Make sure you access the forums. They are probably the most valuable resource that CompuServe has: people who have answers to your questions."

Invest in a map or navigation chart. These maps generally show all of the sections and subsections of the online resources and the routes between them. Maps cost only a few dollars and allow you to consult the big picture before each online session. It will save you time and money.

Use software to preplan sessions. Put in the needed time up front to set up a navigation or scripting program that allows you to plan your online sessions ahead of time. That way the software accesses the online network and performs all operations automatically, again saving you time and money.

Be aware of the virus danger. It's unfortunate, but not all online information is good for you. When you transfer data from the online service to your personal computer, destructive programming code—called a virus—can slip into your system and wreak havoc. Buy the current version of an antivirus program that detects and destroys these unwanted invaders.

Use an "information sifter" to automatically integrate information. Once you retrieve information from online resources, the next problem you have is what to do with all of it. Rather than manually sorting it yourself, use a personal information manager to sort it for you. *Agenda,* for example, includes an Information Sifter capability that acts on all information from external sources such as online data retrieval, CD-ROMs, and electronic mail. It automatically integrates the incoming information into your big-picture categories. Staying current is no problem despite the information explosion when technology finds what you need and automatically organizes it for you.

Find out what's available and how to use it. Many people subscribe to a service, then never use most of the resources. It's like driving a car and never using the radio or climate controls. There's plenty of available help, both formal and informal, to teach you how to make the most of online services. Plugging into online resources can be one of the best ways to connect with a world of contacts, resources, and ideas, and, ultimately, to see the big picture. Isn't that worth a few hours of your time as an investment in your future productivity?

Let others find the facts. If you cannot afford the learning curve to go online yourself right now, there are services that will find the information for you. Burrelle's *Information Search Service,* for example, can tap into some 300 databases, searching as far back as the early 1970s to locate exactly the answers you need and put them on your desk in a matter of minutes. You could even turn to a public library, because many now offer fee-based services. They'll do online searching for you, dig up and copy special documents, or fill any other information need for an hourly fee. Check *The FISCAL Directory of Fee-Based Information Services in Libraries* for participating libraries.

Do you need copies of magazine articles but are too busy to go to the library? The Michigan Information Transfer Source will photocopy and send you an article from just about any business publication (other than local ones) for a $10 fee. You could also contact an information service specializing in secondary research (as opposed to primary research services that conduct surveys, focus groups, etc.). We sometimes use Sharp Information Research on the West Coast for information needs that exceed our time resources.

Step 3—Improve Your Broadband Thinking

Improve your broadband thinking abilities by looking for the GLOBAL picture:

Generate ideas without judgment. Passing judgment on ideas as you make them limits your thinking. It stifles creativity, because evaluation and criticism call for detailed, analytical thinking—the opposite of what you need to be creative. Removing the critic in you unleashes your broadband thinking ability.

- Participate in brainstorming sessions with colleagues and associates to stimulate synergy.
- Think about "why" and "what," not "how to." Why would a customer remain loyal to your firm? What would ensure customer loyalty? Deal with how to accomplish the specifics later.
- Suspend criticism until later when you can afford to be analytical. Don't focus on specifics.
- Attend conferences and professional meetings. Get into thoughtful discussions about issues important to your business.
- Listen attentively to the ideas of others, then respond by building on their ideas.

Link diverse ideas, people, and technology.   Linking is a matter of seeing potentially productive links between dissimilar entities. The process of looking for the links fuels creative and productive thought.

• Read broadly and look for any implications or connections to your field. Buckminster Fuller invented the geodesic dome after reading about how the fly's eye is structured.

• Attend conventions and trade shows. Chat with other attendees who have interesting ideas. Exchange cards and introduce them to others. At a recent conference, several people looked me up and requested meetings at the suggestion of contacts I first interacted with the previous day. The meetings turned out to be mutually beneficial.

• Take a global view in building technology, and link pieces together, if necessary, from far-flung sources. Consultant Michael Sekora gives the example of Japan's Casio Computer, a company that spends "very little money on R&D." But they are very good at identifying a market need and then shopping the world for pieces of technology—some from Japan, more from the United States, a piece from France. "They pull it together, they get in the marketplace, they satisfy the need, and they understand the evolution of the technology so that they can keep on the leading edge."

Open your mind to alternatives.   Keeping an open mind promotes cognitive flexibility. Rather than jumping to conclusions:

• Challenge your own assumptions. Move out of your comfort zone. Who says you can't team with a competitor, as Apple and IBM are doing? Why can't you turn to customers themselves to help serve other customers as Borland and other companies are doing?

• Look for different points of view. If you are a liberal, read what the conservatives are up to.

• Look for new trends. If you can spot them before the media brings them to everyone's attention, you could gain a strategic advantage over your competition.

• Learn something about a new topic every week. You could read about new scientific discoveries, political developments, or economic issues. I like to read *Omni* magazine, because it gives me a glimpse of the latest developments in science and technology yet is an easy read.

- Scan your mail, even your so-called junk mail. There may be gems of opportunity there. You could notice trends, see what your competitors are up to, or spot information to pass along to a colleague.

Blend the new with the old. Isolated facts and figures serve as barriers to knowledge, because nothing exists in a vacuum. Take something out of context, and the information could do you more harm than good. Suppose you want to buy stock in a company. Simply knowing the bottom line—the profit figures—for the company doesn't tell you much about what's going on. You need to see profit and loss figures in the context of the bigger picture. How do profits of this company compare to other companies in the industry? How do they compare with overall economic trends? What about the company's internal and external contexts, what's happening there? You need to see how the latest information relates to what came before to evaluate that information.

- Consider the context. When you encounter a new fact or piece of information, see it in relation to the previous context.
- Avoid drawing any conclusions from information presented in isolation or out of context.
- Pay attention to political and social cartoons. They make you notice what you didn't expect to see by presenting old information in a new way.
- Relate new information to something familiar. My local waterworks did this by using an analogy to allay fears about insignificant amounts of undesirables in the drinking water. "To put these units of measurement into perspective, consider that one part per million in time would be one second out of 11 ½ days and one part per billion in time would be one second out of nearly 33 years."

Associate with a diversity of contacts. In this decade, 14 or 15 million new jobs will be created, and women will fill two thirds of them, according to *Megatrends 2000.* Some 75 to 80 percent of the growth in the American workforce will come from females and minorities. Take a look at your current circle of contacts. Are they representative of these shifts in the workforce?

- Evaluate your colleagues and business contacts. Are they helping you grow and keep up with what's happening today? If not, enlarge your circle to reflect worker diversity.
- Make people the foundation of your effort to *survive information overload.* Develop a varied network of contacts with whom you can share information.
- "Get outside of your own expertise," advises one futurist. The more of a specialist or expert you are, the more important it is to be aware of what's happening *outside* your field. That's why it's important to include people in your network who are so-called outsiders and to read broadly.
- Foster a close work association with someone different from yourself, especially if you represent the majority. If you're male, make it a habit to discuss matters with a female colleague. If you are older, meet with the younger set. If you are not a member of an ethnic minority, make your lunch partner someone who is, then listen to a new perspective on the issues. If you have 20/20 vision and jog for exercise, recruit a physically challenged co-worker to participate on your project or team. You get the idea.
- Expand your group of friends to include a representative and balanced mix of people. Listen to their feedback on your biases so you won't be like the fish in water that doesn't know it's wet.

Look ahead to the consequences.   Develop your ability to exercise future vision:

- Put yourself in your customers' shoes. What do you like about your own company and what you're doing? What needs to be changed?
- Read nonfiction books about future trends in business and society: *Megatrends 2000, 2020 Vision, The Coming Global Boom, Workplace 2000,* and *Managing the Future,* to name a few.
- Ask "what if" questions and consider the impact of the answers on your business.

- Hire someone else to view the future big picture for you. There are futurists and consulting firms that specialize in helping you assess the future. One such firm, Inferential Focus, specializes in inferring important changes that could impact your business long before the media notice any emerging trend or opportunity. The partners of the firm read widely, covering about 350 publications altogether. They look for specific events that stand out as different or unexplained, what they call "anomalous facts," which they discuss as a group on a regular basis. "After we have the specifics," partner Kenneth Hey told me, "we try to step back from those specifics and draw connections across disciplines to see if this is going on in a wider spectrum of events. Our work starts with the details and builds. In that sense, it's the opposite of analysis." This firm's method helped clients anticipate the importance and shortage of water long before it was front-page news.

SELF-TEST: Do You Have a View of the Big Picture?

	Yes	No
1. Do you believe that brainstorming sessions with colleagues are a waste of time?	____	____
2. Are you too busy with day-to-day demands to study future trends and ask "what if" questions?	____	____
3. In the last week, did you scan a magazine or other publication that was totally outside your area?	____	____
4. In the last month, did you use online aids to find some data or information you needed?	____	____
5. Do you regularly participate in a network of contacts that includes a diversity of people different from yourself?	____	____
6. Do you generally handle customer calls when possible rather than pass them off to customer service?	____	____

7. Do you map complex projects with special
   forms in your MCS or with project man-
   agement software?   ____   ____
8. Do you keep an association directory or
   national phone directory on your reference
   shelf?   ____   ____
9. Do projects generally take much longer
   than planned?   ____   ____
10. Do project team members make changes
    in midstream, oblivious to the slowdown
    it will cause?

Scoring

Questions 1–2, 9–10: 1 point for each "yes."

Questions 3–8: 1 point for each "no."

What Your Score Means

Score of 0   Great job! Your broadband thinking abilities
                  are suitable for the information age. Skim the
                  chapter for any new ideas on using online
                  services and anticipating future trends, then
                  go on to the next chapter.

Score of 1–3   You have some perspective on the big
                  picture but could do even better. Read this
                  chapter and follow the Action Plan carefully.

Score of 4–6   You are not paying as much attention to the
                  big picture as you need to in the age of
                  information. Focusing too much on specifics
                  will get you in trouble. Study this chapter
                  and carefully follow the guidelines in the
                  Action Plan.

Score of 7–10   You are too focused on the specifics of your
                  work and need to develop your broadband
                  thinking to survive information overload.
                  Study this chapter thoroughly, then follow
                  the suggestions in the Action Plan. Also,
                  invest extra time in Chapter 7.

Coming up. Once you've developed your broadband thinking, you will begin to see work demands in their proper perspective. One of the biggest demands on a manager's time is the business meeting. The average manager spends nearly half of the day in meetings, talking about work rather than doing it. Dispense with meeting waste and learn to work collaboratively with others to "illuminate the issues." Find out how in the next chapter.

Chapter 5

Too Many Meetings? . . .

Illuminate the Issues (with Graphics and Collaborative Work)

Meetings are the biggest enemy and also the biggest hope to survive information overload.

Everyone bad-mouths meetings. A popular poster sums up current thinking: "A meeting is an event at which minutes are kept and hours are lost." Top executives polled nationwide said that meetings wasted more time than phone calls, paperwork, and travel, according to *The Wall Street Journal*. Survey after survey reports that meetings consume more time and produce fewer results than just about any other business activity.

Yet meetings matter now more than ever. As decisions become increasingly complex in today's information environment, they require a wider variety of expertise. Meetings bring people together to share ideas and forge new and better solutions to complex problems—a true "meeting of the minds."

Those who survive information overload will be those who convert business meetings from boring time wasters to productive, collaborative work sessions.

"We Can't Go On Meeting Like This"

According to recent studies, the typical meeting in corporate America lasts an hour and a half, includes an average of nine attendees, and is called with only two hours' notice (if not regularly scheduled). Most meetings cover the entire agenda only half the time.

The problem of overload caused by meetings becomes more obvious when you look at the percentage of time individuals spend in meetings. Nearly *half* of the average manager's time is spent in meetings (when phone calls are counted as "meetings"). They consume the majority of the workday for senior managers and executives. As one executive put it, "My life is a long string of meetings held together by coffee breaks and lunches." The statistics are staggering:

- 46 percent of the average manager's time goes to meetings.
- Senior managers spend over 60 percent of their time in meetings.
- Meetings take up a whopping 94 percent of an executive's day (counting phone calls and other communications as meetings).

With so much time spent in meetings, little time is left for decision making, thinking, and other intellectual work. On the average, white-collar workers spend a paltry 8 percent of their time thinking and problem solving!

Does it seem that you spend more time doing productive work and less time in meetings than these averages suggest? If so, consider the following fact.

Nearly 90 percent of us *under*estimate the time spent in meetings and *over*estimate the time spent on intellectual work.

In all likelihood, you spend more time in meetings than you realize, especially when you consider that every phone call is actually a meeting between the two (or more) parties.

Corporate Meeting Waste

A nationwide survey by Accountemps found that executives waste up to three and a half months of the work year on distractions, primarily in the form of unnecessary meetings. To find out how much time was wasted in meetings at their company, a group of managers at McDonnell Douglas surveyed 100 of the company's internal meetings. One out of every five meetings was found to be completely unsatisfactory: objectives were not met, participants were ill-prepared, and no summary of the meeting was provided. The *majority* of the meetings had no agenda, no secretary/ recorder, and no minutes. Many did not begin or end on time. Nearly a third had no objective—no stated reason for having the meeting in the first place!

Based on these statistics, the study group concluded that 185,000 hours per year were wasted in meetings. With the average manager's time worth about $50 per hour, that amounts to over *$9 million of waste* in one company alone. The annual meeting waste in the United States adds up to a whopping *$37 billion* primarily in wasted wages, according to the *Washington Post*.

The reasons that most meetings waste time include:

- No specific, clear-cut objective for the meeting, its leaders, or its participants.
- No meeting agenda.
- Too many or the wrong choice of participants.
- Inability to present ideas concisely.
- Improper use of visual aids.
- Too many digressions and interruptions.

Furthermore, a survey by the University of Southern California found that when 903 managers at 36 companies were asked about their last meeting, one third of the managers said they had no impact on the final decision that was made. One third felt pressured into supporting a position they disagreed with. Clearly, meetings have not lived up to their potential for the majority of managers.

How Much Do *Your* Meetings Cost?

You can calculate roughly what your meetings cost, and the figures may surprise you. As an example, assume that five employees each making $50,000 annually attend a weekly meeting lasting two hours. In one year, that weekly meeting consumes $12,480 worth of resources in the form of personnel costs alone. Of course, many other factors contribute to meeting costs. In addition to an employee's base pay, add taxes, benefits, overhead, audiovisual needs, and possibly travel time and expense.

Here's another example: What do you think it costs for a two-hour meeting attended by a manager earning $80,000 a year, four managers at $60,000, and six staff members who average about $40,000 annually? The cost amounts to over a thousand dollars—good reason to give careful consideration to both the necessity of the meeting and who should attend.

Personnel Costs Per Hour

	Number of Participants						
Average Pay	4	5	6	7	8	9	10
$80,000	$308	$385	$461	$538	$615	$692	$769
70,000	269	337	404	471	538	606	673
60,000	231	289	346	404	462	513	577
50,000	192	240	288	337	385	433	481
40,000	152	192	231	269	308	346	385
30,000	115	144	173	202	231	260	288

Source: Minnesota Western.

Calculate the cost of your last meeting using the accompanying table. The figures represent fully capitalized costs per employee hour. Then ask yourself, "Could a memo have produced the same result as the meeting for a fraction of the cost?" Sometimes a conference call will suffice.

When you view a meeting as an investment that must show a return, planning the meeting warrants careful consideration. Consider substituting another form of communication such as a letter, memo, or conference call to make sure the payoff of holding the meeting justifies the investment of staff time and other resources.

The Wharton Study and Beyond

Researchers have discovered a way to shorten meetings considerably and make them more effective as well. Meetings that drag on with little accomplished are due in part to an overabundance of verbiage, according to a six-month study conducted at the Wharton Business School under a grant from the 3M Corporation. The study simulated 36 business meetings under three controlled conditions. In one, the presenter used graphics (overhead transparencies) to argue in favor of the introduction of a new product ("Crystal" beer), while the presenter arguing against it did not use meeting graphics. In another, these roles were reversed, with the presenter who argued against the product using the overhead transparencies. In the third, neither presenter used overhead transparencies. The findings supported the use of visual aids in meetings.

Graphics shorten meetings. The meeting time was reduced by an impressive 28 percent when presentation graphics were used. The meetings with overheads generally lasted 18.6 minutes, compared with 25.8 minutes for meetings without overheads. The use of overheads resulted in less time spent on lengthy monologues and more time spent on interaction. According to the study, "the less time spent in monologue, the more efficient the meeting, because more real interaction of ideas is allowed to take place."

Graphics are more persuasive. Under the first two experimental conditions, the presenters who used graphics succeeded in persuading a group in their favor 67 percent of the time, while the verbal presenters succeeded only 33 percent of the time. The study was set up to test the effectiveness of graphics whether the graphics presenter was arguing for or against the proposition (introducing a new product). When overheads were used to promote a "go" position, the majority agreed with the "go" position. Similarly, when visuals were used to promote a "no go" position, the majority agreed with the "no go" position. When no overheads were used, there was a 50–50 chance that the group would agree with either presenter.

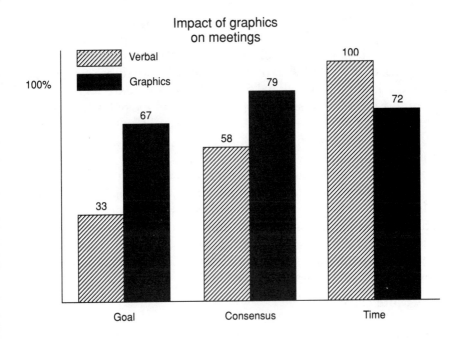

Impact of graphics
on meetings

Graphics speed decision making. When graphics were used in the meeting, 66 percent of the attendees said they made their business decisions "immediately after the visual presentation." When graphics were not used, most of the attendees delayed making a decision until "sometime after the group discussion following the presentation." It's no wonder meeting time is longer for meetings without overhead support—everyone needs to discuss the issue longer to make a decision.

Graphics promote group consensus. The goal at each meeting was to reach a group consensus within the time allotted. Reaching a consensus occurred in 79 percent of the groups that saw graphics, compared to a 58 percent consensus rate for the groups that were not shown the graphics. A failure to achieve consensus occurred in 42 percent of the groups not exposed to the overheads, double the failure rate for groups who saw the overheads.

Graphics make a good impression. The presenters who used graphics in the meetings were perceived to be more effective than those who did not use graphics. Specifically, presenters who used the overhead transparencies came across as "better prepared, more professional, more persuasive, more highly credible, and more interesting." The researchers looked at attendees' responses to 11 key presentation characteristics, including strength, conciseness, professionalism, clarity, precision, commitment, interest, and attractiveness. On every one of the 11 variables, the presenter using transparencies fared better.

In a follow-up study also funded by 3M, the University of Minnesota tested the effectiveness of visual aids in the form of 35mm slides as well as overhead transparencies. The study found that presentations using visual aids were 43 percent more persuasive than unaided presentations. Like the Wharton study, presenters using visuals were perceived as being:

- More concise.
- More professional.
- Clearer.
- More persuasive.
- More interesting.
- More effective in the use of supporting data.

The University of Minnesota/3M study went further in some respects than did the Wharton study. Graphics were found to result in better attention, comprehension, and retention. Color graphics were more persuasive than black and white, especially as a means of inciting meeting participants to action.

Meeting Graphics—What Medium to Use?

You would think that with all the proven benefits, graphics would be used extensively in typical business meetings. Right? They're not, according to an Annenberg School of Communications survey of a variety of companies of different sizes and lines of business. In 47 percent of meetings, graphics were not used at all, while in 44 percent they were used less than half the time. In only 9

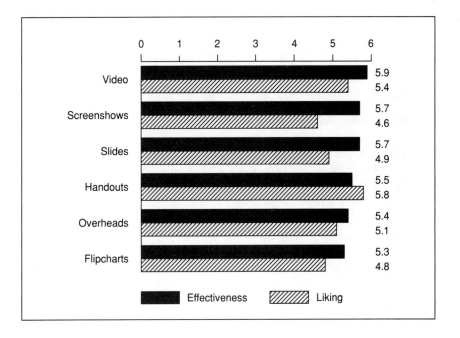

percent of the meetings were graphics used more than half the time.

The study also looked at the effectiveness of various media (slides, overheads, printed handouts, etc.) and how well attendees liked each medium. The most popular medium, according to attendee ratings, was the printed handout, but video was judged to be most effective. Handouts were used at 47 percent of the meetings; overheads and chalkboards were the second most popular, at 13 percent each.

In the past few years, I have written several guides full of tips on what type of media to use in meetings and how to use media effectively (*Master Graphics: Effective Overheads for Business Presentations* and *Power Graphics: Your Guide to Intelligent Presentations*). Here's what you need to know in a nutshell about selecting the right medium for your meeting:

- Use overhead transparencies for informal meetings. They're also a good choice for persuasive communication, because you can easily maintain eye contact with attendees.

Box 5–1

The Business Meeting: How CEOs See It

• Uses an electronic meeting system to bring all departments together to create a comprehensive strategic plan. "We feel there is no other way to develop the same quantity of quality output from 30 individuals in a reasonable amount of time."

Samuel Eichenfield, *CEO, Greyhound Financial*

• Promotes group decision making in meetings. She recounts a case in point: "I laid out the pros and cons and allowed the team to come to the decision collectively."

Sandra Kurtzig, *CEO, ASK Computer Systems*

• Has weekly meetings with key subordinates who have planned the three or four points they want to make.

Andrew Grove, *CEO, Intel*

• Likes meetings so he can talk face-to-face. "I want to look at a person's eyes when he's telling me something."

Robert Crandall, *CEO, American Airlines*

• Adjourns meetings on time and reconvenes, if needed, during spare time. "You'll see a lot of meetings that convene at six when people can take time to really think through a problem."

Ryal Poppa, *CEO, Storage Technology*

• Considers personal information manager (PIM) software such an integral business tool that it is a prerequisite to all meetings between him and his managers. If they come without it, "I tell them, 'I don't think I can talk to you now. Get your laptop.' "

Randy Fields, *Chairman, Mrs. Fields*

- Plan to use overheads for international meetings, because overhead projectors are available worldwide.
- Choose 35mm slides for a formal presentation or if you need to cover a lot of material in a short time.
- Begin a formal meeting by delivering the facts with 35mm slides to establish a credible, professional image. Then use video to persuade and recruit support.
- Hand out printed copies of information if there is a lot of specific data that needs to be studied. Make sure hardcopy prints are annotated and self-explanatory. You can include more detail in printed output than in projected media such as slides.
- Show videotapes to hold attention and to motivate people to take action in a sales meeting or other decision-making meeting.

How to Make Meetings Matter: *Collaborative Work*

The key to make meetings matter is *collaborative work* among the participants. In order to convert meeting waste to productivity, participants must take an active, rather than passive, role.

You can foster collaborative work by making the Master Control System a part of every meeting. It allows participants to integrate notes and assignments with their other activities. Attendees contribute fully to the discussion because they're looking at the big picture. Follow-up tasks get done when they are planned and scheduled during the meeting, because each person incorporates the task into his or her schedule.

Standardizing the MCS across the group contributes to collaborative work. In other words, it helps if all staff members use the same format. When everyone uses the same format, language, and agenda, communication flows more easily. One example format is the Business Meeting Checklist from Time/Design. The form is designed to allow participants to keep their focus on agenda items, subjects discussed, decisions and conclusions reached, and who is accountable for any follow-up action.

Miles Babcock, marketing director for Time/Design, told me how helpful the checklist has been at company meetings. "I've

Source: Illustration courtesy of Time/Design.

found that I can hold people extraordinarily accountable. They know that at the next meeting, everybody there is going to know who was given the responsibility and will expect a complete status report of how it went. 'It should be done by now, Bill, so how did it go?' Another benefit of the form is that you create your own agenda as well as what may have been distributed to you prior to the meeting." During an interdepartmental meeting on one issue, the training director or customer service manager may raise another pressing issue related to their own agendas. "If I hadn't informed customer service that marketing mailed the yearly refill promotion, they might be inundated suddenly with phone calls and not know why. It helps to keep the right hand constantly aware of what the left hand is doing."

One of the two people I've worked with who is best at running a meeting is venture capitalist Jim Treleaven, formerly the general manager at a high-tech company. He kept track of important facts reported at the meeting, what was to be done after the meeting, and who was responsible by making a few notations in his planner. Then he followed up to make sure it happened. The other master of meetings is Dr. Patricia Marshall, who directs a busy public television station. Last time we met, she called in several

Subject/Theme			Business Meeting Checklist

Date   DataBank#
Start   Finish
Agenda

		OK	Participants	OK
1				
2				
3				
4				
5				
6				
7				
8				
9				
10				

Notes

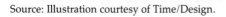

Source: Illustration courtesy of Time/Design.

staff members to answer questions and resolve issues on the spot so projects could move forward. These two people know the secret of effective meetings: collaboration and follow-up.

Meetings will matter when they engender a true *collaboration*—the watchword of the 90s and beyond. For a large organization to thrive today and in the future, collaborative work must become a reality. The first step is cross-functional communication. Sales must talk with marketing, shipping must interact with the credit department. The easiest way to take that first step is to reach for the phone.

Collaborative Meetings by Phone

Every business phone conversation can be thought of as a meeting—a chance to link up with other departments, business associates, suppliers, and customers. Too often, however, callers get caught in an unproductive game of telephone tag. A basic resource to help you break out of that rut is George Walther's book, *Phone Power*. He recommends, for example, that you ask the secretary to page an unavailable party, ask for another number where the person can be reached, or schedule the call like you would a face-to-face meeting.

The aim is to convert your telephone from costly nuisance to productive tool by following these three guidelines.

1. "Conference in" the big picture. It's easy to get off on tangents while talking on the phone. With so many other things on your mind, you can get off the track, forget key issues to discuss, and lose track of important points you wanted to make.

- As a general rule, do not make or take a phone call without your MCS in hand (or contact management technology discussed subsequently). Use it to keep track of the discussion and outcomes for better follow-up.

- Glance at the first page of the related category section in your MCS before making a business call. It will remind you of how that particular call fits into the bigger picture, and you won't waste time getting off on tangents.

- Write down the one or two key points you want to make or issues to resolve during the call. Check them off as you cover them during the phone call.

"Can you hold for a moment? I'd like to conference in the right side of my brain, if you don't mind!"

© 1991 Brad Veley.

- Ask questions related to the objectives if the discussion gets off track. "So now, when will you send that report?"
- Summarize what is being accomplished during the call in relation to the goals or big picture. Do the same before hanging up. "You're going to finish your part by Friday so we can wrap up this project on time."

2. Leverage your time with phone technology. Most likely, you are well aware of the time-saving benefits of voice mail and messaging systems. The following are several additional ways you may benefit from telecommunications.

- Use a personal voice telegram service such as the *AT&T Message Service* or *Allnet Call Delivery* when you can't get through to someone and time is short. You simply call the service and leave a message up to one minute in length. The service then charges you $2.50 or so to deliver your message to a particular person or somewhat less to automatically call the phone number and deliver the message at the time you specify. The recipient may leave a one-minute reply. Your office computer system can also be set up to send voice telegrams.

- Make commute time productive with a cellular phone. According to a Gallup survey, there will be 33 million cellular-phone users by the year 2000. More than half of the 5.7 million users today say that a car phone has increased their business revenues and productivity.
- Set up a three-way call or teleconference to resolve issues quickly or disseminate rapidly changing information. It is possible to have over 500 conferees, which is feasible as long as most participants are listening only.
- Consider a pager. The newest models allow you to receive text messages sent from a computer, alert you when an important call comes in, and can receive group pages. Two-way wireless satellite communication on a wristwatch, a la Flash Gordon, is no longer far-fetched science fiction.

3. Automate the contact process. Contact management technology (see the resource directory for specifics) helps individuals or groups synthesize details related to people contacted by phone or mail. The technology brings automation to labor-intensive fields such as sales and marketing. Automation helps you keep track of people to call, follow-up activities, and next steps.

- Keep the phone numbers of people you need to contact in a contact database with auto-dialer. That way you'll save time and expense by letting your computer correctly and quickly dial numbers while you focus on your objectives.
- Choose a contact manager with a pop-up notepad so you can take notes during the conversation.
- Add form letters and other follow-up documents to the contact manager file. Then select the appropriate documents for follow-up during the phone meeting.
- Better yet, add a fax board to your system and fax follow-up material before you go on to the next call. It's a sure way to keep follow-up letters from falling through the cracks.
- Scan brochures and other documents containing graphics or signatures. Again, store these in your contact manager so you can transmit them with the touch of a key. Otherwise

the request for follow-up material adds to the workload, piling up in an in-basket.

- Network with others in your organization to coordinate contact management efforts. Integrated systems are available that automate correspondence and many other tasks. Check the resource directory for systems for corporate-wide sales and marketing automation.

The telephone is the manager's basic tool for collaborative work. The ultimate productivity tool, however, is a combination of this technology with two others: television and computers.

Knowledge Navigator: **A View of the Future**

Apple Computer conveys its vision of the ultimate productivity tool in *Knowledge Navigator*. This short video depicts a college professor as he prepares for a lecture about deforestation in the Amazon rain forest. Using a futuristic computer-TV-telephone no larger than a notebook, he calls up his colleague in another state. She appears on his screen, and they begin a videoconference. Unlike today's videoconference, however, she simultaneously sends a visual simulation of her latest data on the spread of the Sahara over the last 20 years that instantly appears in the center of his computer screen. He directs the computer to put her Sahara simulation together with his simulation of how current logging rates will affect the Amazon. They both watch the combined simulations and witness the Sahara desert grow as the rain forest decreases in area. They then so some "what if" scenarios in which they decrease the logging rate in the rain forest to see its simulated impact on the desert. They are satisfied with the results, and each makes a copy of the new simulation data before ending the productive, yet brief, session.

When I show this video during seminars, viewers are impressed, but many don't initially grasp the full significance of the future vision. The implication of *Knowledge Navigator* is that video-tele-computer-conferencing will be *more* productive than traditional face-to-face meetings. The reason is that each participant can share simulations, demonstrations, databases, spreadsheets,

Knowledge Navigator.
© 1987 Apple Computer, Inc. Used with permission.

photos, film clips, and other materials. Meetings would not be the largely passive, verbal experiences they are today but would become interactive work sessions supported by incredibly capable tools. Meeting around a table with nothing more than a pad of paper or a planner will be as unthinkable in the future as today's surgeon trying to do an operation without a fully equipped hospital operating room containing millions of dollars in technology to support the effort.

Five Technologies for Collaborative Work

Current technology has not achieved quite the level of *Knowledge Navigator*, but many similar capabilities are possible. Computers, telecommunications, and facsimile technology offer opportunities for people to work together in closer collaboration.

1. Groupware for collaborative computing. There is software available today that allows people to work together in dynamic collaboration. The computer equivalent of a conference call, it's called "collaborative computing" or "groupware." It allows participants in different locations to share information, including all types of computer files. A person in California can connect with colleagues in New York, Japan, and Europe to work together simultaneously on the same computer document. Each person can change or add material—a word, paragraph, graph or other visual—while seeing what the others are doing.

Collaborative computing software is saving time and money for companies, especially those with worldwide operations. "If we want to hold a two-hour meeting and bring together people from Europe, the Orient, Canada, and South America, we will lose four days for each person," explained Neil Drake of Dow Chemical. Collaborative computing "lets us hold a two-hour meeting in two-hours." The computer meeting requires no travel and may be more productive since each participant can pull in and share data that may prove important as the meeting unfolds. One user of this technology summed up the process as "more of a dialog than a set of soliloquies."

"We think there's a dynamic energy or synergy that really gets going when people work together," said T. Reid Lewis, president of a firm that has created groupware called *Aspects*. "If we're talking together, I get myself all psyched up and focus all my attention on one particular problem." The software lets a number of people work simultaneously on the same document. Each person signs on and picks a particular symbol to use as a pointer. The meeting begins once the person who organized the meeting distributes an electronic copy of the document to each participant. The organizer determines whether the meeting will be free-wheeling or if each person needs to get permission to make changes. People who "talk" too much can be "shut up" more easily than in a face-to-face meeting. If the meeting is free-flowing, any user can make changes that show up almost immediately on the screens of the other users.

When I asked Lewis how collaborative computing alleviates information overload, he explained that, for one thing, it helps you quickly get what you need from busy people such as experts. "As more information builds up, people get more specialized. You can get an expert involved without a major expense or hassle. They can look at your document if it's a contract, a scientific treatise, or maybe just a proposal. You can call on the expert and say, 'Look at this and tell me what you think.' You get the feedback in context so if they say 'I think you mean this' and you don't, you can immediately say, 'That's not what I meant; here's what I

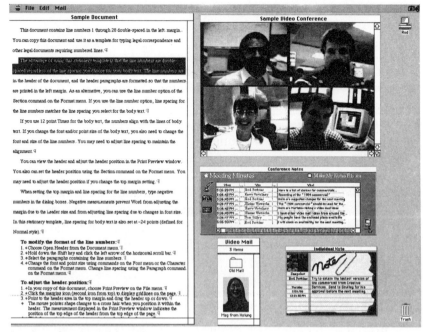

A screen from Apple's in-house, experimental videoconferencing system.
Source: Illustration courtesy of Apple Computer, Inc.

IBM's TeamFocus room for collaborative computing.
Source: Illustration courtesy of IBM.

want to accomplish.' " It's almost like a face-to-face meeting where you can point to sections during the interaction. That's hard to do over the phone.

2. Electronic meeting rooms for collaborative computing. Participants need not be in far-flung locations to take advantage of collaborative computing. A 1980s project called CoLab at the Xerox Palo Alto Research Center, for example, allowed managers to work together in one room and share information from individual computers on a community screen referred to as a "liveboard." Today, other companies are adapting the idea to current technology. Electronic Data Systems set up a conference room to facilitate small-group meetings. Called the Capture Lab, the meeting room features eight personal computers around an oval table with a large community screen at one end. Participants come together, work privately and/or publicly, share documents (spreadsheets, databases, graphics, and so on) with the group, make changes in those documents, and leave with the updated information to continue individual work efforts.

There are several commercially-available systems for electronic meetings. According to users of these systems, electronic meeting systems are very effective. Meeting time is reduced by 40 percent

The "Capture Lab" at EDS.
Source: Illustration courtesy of EDS, Center for Machine Intelligence

or more. Group brainstorming is more productive. Decisions and group consensus occurs more quickly as a result of electronic rating and voting. Electronic meeting rooms can be rented or set up temporarily if your group cannot bring this technology in-house.

3. Voice-fax-computer technology. Not long ago, I enjoyed the benefits of a new breed of "full service" customer support thanks to the combination of voice, fax, and computer technology. I called a major software company to ask questions about their product since I was considering an upgrade. My questions for our "phone meeting" were: How does the upgrade differ from the current version? Would the upgrade also run on a laptop computer? How much space would it require on the hard disk? A call to technical support put me in touch with the technical rep, who answered the third question, then switched me to an automated information retrieval system. With a few keypresses on my phonepad, I ordered reports to answer the other two questions, keyed in my fax number, and hung up. A few minutes later the reports were printed out on my fax machine, and I had the needed answers.

When you consider why you meet with others, especially by phone, the answer is often that you want to exchange information—to ask questions and get the answers. For those phone meetings that are primarily or exclusively for the purpose of question and answer, "combination" technology provides a cost-effective alternative to both phone and face-to-face meetings.

4. Online networks, forums, and roundtables. In its simplest terms, an online computer meeting is essentially an E-mail session taken one step further: there's a simultaneous, two-way exchange. It's something like a phone call, except "callers" type their comments back and forth rather than talk. There are no preset limits on how many people can participate at once, but like a traditional meeting, an online meeting can bog down with too many people.

To participate in online meetings, you first access a network. Computer networks are akin to telephone systems in that some are public and some are private. The difference is that more computer networks are private, internal company systems that don't interconnect with other outside networks. That arrangement is just about as useful as a phone system that only lets you call others within your company. Many companies are moving quickly to interconnect with other networks to get the full benefit of computer networking.

As more companies interconnect, online meetings are catching on and dramatically changing meeting dynamics. The whole issue of status is different in an electronic meeting. "In face-to-face groups," explains one corporate psychologist, "the person with the highest status tends to dominate, whether that status was earned or not. In an electronic group, the effects of status are reduced." Others pay attention to what someone contributes, not who they are.

Compared with a traditional meeting, there is not much you cannot do in an online meeting and many new things you can do. As in a face-to-face meeting, you can hold a private conference online, find out who else is "around the conference table," and contact someone who is busy in another electronic meeting. In addition, you can share computer files such as spreadsheet data, reports, or programming code. You can meet with people from around the world without leaving your office. Perhaps best of all, you can learn from other people's experience.

Electronic meeting groups are organized into "forums" on CompuServe, and each of the several hundred forums maintains a storehouse of information that builds up over time. Stewart Cheifet, host of the PBS weekly show *Computer Chronicles,* says that he uses the forums to get information about new products. "For example, I was interested in buying a camcorder, so I accessed the Consumer Electronics Forum and found at least nine months' worth of product recommendations from other users. You just can't get that kind of information anywhere else."

More flexible than traditional meetings, online meetings let you meet simultaneously or allow individual participants to interact at times most convenient to each. You can ask a question or make a comment one day, then check back the next to get feedback. As you can see, there is a fine line between online meetings and E-mail. Both allow you to leverage your time and effort by connecting with others to plug into the collective intelligence of the group.

5. Videoconferences and satellite communications. We haven't reached the point of the futuristic *Knowledge Navigator* yet, but videoconferencing is playing a role in group meetings, especially meetings that would otherwise require travel.

- Managers at the IBM plant in Austin, Texas, have two videoconference rooms to save time and money on airplane trips.
- Sam Walton, founder of Wal-Mart, talks to all of his 275,000 employees weekly over the company's satellite network.
- Alaska's firefighters, who are dispersed across a wide area, participated in a government-sponsored videoconference that allowed interaction via toll-free phone response. I've conducted several seminars this way, and the setup is an effective, low-cost alternative to two-way videoconferencing.

ACTION PLAN: Illustrate the Issues in Better Meetings

Meetings are a must as companies move from national to international, hierarchical to horizontal, and homogeneous to heterogeneous structures. As a result of globalization, diversity among workers has increased, while their ability to share a common context and language has decreased. The need to see the big picture has skyrocketed. The challenge is to communicate the big picture in concrete terms that everyone can understand.

And don't overlook the fact that meetings are important to your career. Four out of five managers evaluate each other based on how they participate in a meeting, according to research done by Harrison Conference Services and Hofstra University. Also, 87 percent judge leadership based on how a person runs a meeting. It pays to master meetings, making them productive tools rather than time wasters that contribute to overload.

Step 1—Eliminate Unnecessary Meetings

Before your next meeting, ask yourself these 10 basic questions before you invest your time and effort:

1. Am I needed as a participant in this meeting, or will I be essentially a spectator?
2. Must I be in attendance to make a contribution?
3. Can I participate in some other way: write a memo, provide data, make comments over a speaker phone, or give other input that doesn't require my presence?
4. Can I get the important points of the meeting by getting a verbal report or reading a summary of it?
5. Will the meeting be available on audiotape so I can listen while I commute or during "wait" times?
6. Can I benefit my organization more by doing something else?
7. Can I participate in a conference call, videoconference, or collaborative computing session to achieve the same objectives?
8. Does the group need immediate input from me that I could fax to them in one or two pages?
9. Is the meeting the result of an individual issue that could be handled by an informal, personal conference?
10. Must I attend the entire meeting, or can I selectively attend only the part that is pertinent to me?

Step 2—Keep Necessary Meetings B•R•I•E•F

If your answers from Step 1 do not obviate the need for attendance, then the following five-step plan will help you maximize the meeting's effectiveness yet minimize your effort.

Begin meetings with the big picture.   When you start with the big picture, it sets a decisive tone and puts everyone on notice that this meeting is important in the larger scope of things.

• Begin by communicating the big picture to all participants. It's not enough for the presiding officer to know what is to be accomplished.

- Convey the big picture verbally or, better yet, hand out a simple chart or map. Everyone at the meeting should be able to answer these questions: What is the point of the meeting? How does it fit in with other endeavors? Why is it important?
- Visually convey the main goals of the meeting with a simple chart. Draw it on the board if it's not preprinted.
- Position the most important issues or goals higher on the chart than secondary matters.
- Summarize supporting facts and details needed to make decisions on the major issues and hand these out to participants.

Restrict participants, topics, and time. The tendency is to include everyone from one department or unit. Many issues today are better served by a cross-functional approach. Some members of the department could actually be "outsiders" when it comes to the issues to be discussed, and those people are invariably the ones who end up leading the discussion off track. Generally, the more people at a meeting, the more difficult communication becomes.

- Invite only those people who have a bona fide reason to attend and contribute to the meeting objectives. Limit attendance to those directly involved. Make sure, however, that all interests are represented.
- Distribute an agenda far enough in advance to allow everyone to prepare. A group of CEOs polled nationwide said that one of the top three ways to improve meetings is to send out an agenda along with backup materials *before* the meeting.
- Deal only with topics listed on an agenda that was distributed before the meeting.
- Cover topics not on the agenda, if they are important, under new business. Schedule it at the end *after* meeting goals have been achieved.
- Distribute the agenda at least two days in advance for an informal meeting. Allow a week or more for meetings that require serious preparation.

- Include only a few major points or issues on the agenda. If there are more than several to deal with, schedule a separate meeting for additional topics.
- Be sure that the agenda includes the starting and ending time of the meeting, the location, and who should attend.
- Encourage notetaking on the agenda by leaving space on the sheet for comments so participants can *see* when the dicussion is getting off track.
- Print the agenda on prepunched paper or forms that can be filed in each person's MCS if a critical mass of people use the same system.
- Limit meetings to a reasonable length of time and stop the meeting when the limit is reached. CEO Ryal Poppa told me his company adjourns all meetings after two hours and reconvenes them after 6:00 P.M. People who don't need to stay for the additional discussion are free to leave; those involved don't feel rushed.

Illustrate main points. "What you see, honey, is what you get," proclaimed Geraldine (Flip Wilson's comedic character). Those words apply to meetings, because attendees remember what they see. Research studies confirm that people are up to five times more likely to retain information that is visually rather than verbally presented. It makes sense to visually portray what's most important: the main points.

- Show cartoons to make an important point in an interesting way. Make sure, however, that the cartoon is relevant not merely to the general topic but to the specific point that you want to make.
- Avoid any illustration that is not related to the central points or meeting objective. Such illustrations are commonly included "to add interest and motivate attendees," but they only serve as a distraction and could hinder the meeting.
- Match the medium that you use (35mm slides, overhead transparencies, hard-copy prints, videos, etc.) to both your message and the type of meeting. Overheads are appropriate for small, informal meetings; slides are better

for large, formal presentations. Attendees can study the information on printed hard copy at their leisure; videos pick up the pace and ensure a "standard" presentation, useful for sales presentations and product demos.

- Use each medium effectively. For overhead transparencies, turn off the projector once people have read the information. Change slides every 10–20 seconds. Hand out detailed printed material at the end of the meeting so it's not distracting. Keep video presentations brief and relevant to the point of the meeting.
- Clarify complex issues with a simplified, big-picture chart. I've created one-page charts that helped clarify the issues at both university meetings and meetings with consulting clients. In both cases, others have adopted the charts and used them in subsequent reports, discussions, and follow-up meetings.

Expect objectives to be met. Expect the objectives to be achieved, and they will. Webster defines a business meeting as a gathering of people actively directed towards a specific, immediate goal. When participants know what the goal is, chances are it will be achieved. The key is to communicate the objectives to participants.

- Refer to the big picture during the meeting and gauge progress in relation to it. The big picture should include the objectives of the meeting—that is, what should be accomplished by the end.
- Condense the point of the meeting to the barest essentials.
- Do not include more than several main objectives for any one meeting.
- Use simple wording to describe the objectives and state them in concrete, actionable terms. New jargon or terminology that you plan to introduce during the meeting should not be part of the objective statement.

Follow up with a summary. After the meeting, send participants a summary that includes major decisions made or actions taken. The summary will reinforce participants to take a big-picture view of the situation.

- At the end of the meeting, make sure everyone enters commitments in their Master Control Systems to assure timely follow-up.
- Send out the summary of a regular business meeting to all participants within two days of the meeting.
- Resist any temptation to provide detailed minutes. They're not only a waste of time and paper but they also obscure the central question: What has changed as a result of the meeting? Agreed To, Decided
- Include a record of major announcements, reports, and motions along with the standard details: presiding officer, place of meeting, time and date, and the date of the next meeting.
- Write the deadline and the names of individuals who are to take action next to each major decision on the summary.

SELF-TEST: How Keen Is Your Meeting Savvy?

Take the following self-test to find out whether meetings are making—or breaking—your career and your company.

	Yes	No
1. Do you invest too much time in meetings for too little return?	____	____
2. In the last month, did you attend a meeting that had no specific, clear-cut objective?	____	____
3. In the last month, did you attend a meeting that provided no agenda prior to the meeting?	____	____
4. In the last three months, have you attended a meeting where a key person failed to show up for some nonemergency reason?	____	____
5. In the last three months, did a person not connected with the point of the meeting attend anyway and lead the discussion off track?	____	____

6. In the last three months, did you fail to
    receive the minutes or a summary of a
    meeting?   ____   ____
7. In the last three months, did you receive a
    concrete plan outlining what was expected
    of you at a meeting?   ____   ____
8. Did the most recent meeting you attended
    have a secretary or recorder?   ____   ____
9. Did the most recent meeting you attended
    include charts, graphs, or other images to
    illustrate main points or otherwise aid dis-
    cussion?   ____   ____
10. Did the most recent meeting you attended
    start on time?   ____   ____

Scoring

Questions 1–6: 1 point for each "yes."
Questions 7–10: 1 point for each "no."

What Your Score Means

Score of 0   Great job! You are not wasting time in
                  unproductive meetings. Skim the chapter for
                  any new ideas on alternatives to the face-to-
                  face meeting and what meetings of the future
                  will be like. Then go on to the next chapter.

Score of 1–3   You are managing meetings well but can do
                  better by following the Action Plan in this
                  chapter.

Score of 4–6   You could be significantly more productive if
                  you convert meetings from time wasters to
                  productive tools. Read and study this
                  chapter, then apply the Action Plan to your
                  next meeting.

Score of 7–10   Meetings consume too much of your time
                  and energy with too little return. Read and

study this chapter carefully, then apply the
Action Plan to your next meeting. After the
meeting, reread the Action Plan and reapply.

Coming up.   One reason for so many meetings is that we
must keep up with new information. In the information age,
learning is an ongoing, important activity. It boils down to this: "If
you want to keep earning, you better start learning." More than
ever before, we need to learn how to learn, something that is not
taught well in school. The key is to "visualize new concepts." Turn
to the next chapter to find out how best to learn.

Chapter Six

Too Much to Learn? . . .
Visualize New Concepts
(With Chunking, Visual Analogies,
and Structured Thinking)

The rate at which individuals and organizations learn may become the only sustainable competitive advantage.

A. R. Stata, *Chairman, Analog Devices*

Whether as a company or an individual, it makes good sense— and could mean your survival—to invest in training and education. "Any company that aspires to succeed in the tougher business environment of the 1990s," according to the *Harvard Business Review*, "must first resolve a basic dilemma: success in the marketplace increasingly depends on learning, yet most people don't know how to learn." And, I would add, many companies don't know how to train.

A popular myth is that learning is largely a matter of motivation. If a worker has the right attitude, learning surely occurs. So companies use compensation programs, other incentives, and reviews to motivate workers. The problem is that learning is not simply a matter of commitment.

Increasingly, the key to effective learning in the information era is how you *think*, not how you *feel*.

To remain competitive and survive into the next century, individuals must learn how to learn, and organizations must learn how to train effectively.

A Look at Employee Training Today

Seventy percent of the people who will be working in the year 2000 are on the job now. The majority of them, 75 percent, will need additional training before the turn of the century to keep up with the new demands of their jobs.

- U.S. companies spend $30 billion on formal training each year.
- This figure represents only 1.4 percent of payroll.
- Only 10 percent of employees receive formal training.
- Yet 42 percent of the work force, roughly 50 million workers, will need training in this decade.
- In addition, 37 million more workers need entry-level training each year.

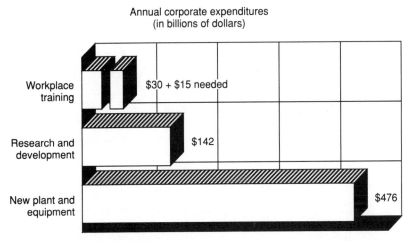

Annual corporate expenditures
(in billions of dollars)

Workplace training $30 + $15 needed

Research and development $142

New plant and equipment $476

To put training expenditures in perspective, consider that companies invest $142 billion in R&D and $475 billion a year for capital improvements to plant and equipment. "Yet it is very well documented that the gains in productivity from workplace learning exceed the gains from capital investments by more than two-to-one," says the president of the American Society for Training and Development (ASTD), John Hurley.

Nearly 50 million employees will not receive the training they need unless companies expand training budgets by 50 percent. The additional training is for workers who need skills and technical training (16 million); executive, management or supervisory training (5.5 million); customer service training (11 million); and basic skills training (17 million). The estimate does *not* include the approximately 37 million workers annually who need entry level or "qualifying" training.

American workers who need training

Training category

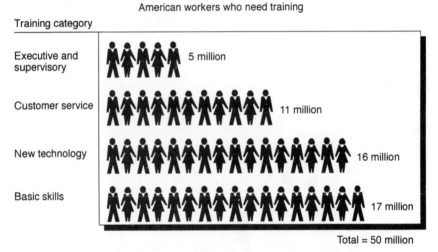

Training category		
Executive and supervisory		5 million
Customer service		11 million
New technology		16 million
Basic skills		17 million

Total = 50 million

© 1990, the American Society for Training and Development, Inc. Reprinted with permission. All rights reserved.

To some extent, institutions of higher education may help fill the gap. Of the $12 billion spent annually on executive development in the United States, about 25 percent goes to business schools. Nearly 10,000 managers work toward master's degrees on weekends at business schools in North America, up 27 percent from four years ago.

Reasons to Invest in Training

In the eyes of some managers, the benefits of training are obvious. "We are expecting training to flourish in the 1990s," says Suzanne

Hovdey, director of corporate communications for Times-Mirror. The reasons include corporate downsizing, the weakness of the U.S. education system, and the rapid rate of change in the workplace, which puts "more attention on building and retaining a qualified work force. Training is a key element in that."

Yet many companies underfund training, leaving a total "training gap" of $15 billion each year in the United States alone. Clearly, the message is not getting through to the right people. Here is a rundown of the major benefits of employee training.

Training yields profits! Productivity gains from training have been well documented and generally produce double the gains from capital investments. One company, Motorola, that has been described as "the company most committed to the task of lifelong learning" retrains one third of its 99,000 employees each year.

"Just as when you buy a piece of capital equipment, you put aside money to maintain that equipment, we require that 1.5 percent of payroll be put aside to maintain the competency level of the employees," says Bill Wiggenhorn, head of Motorola's training.

Why does Motorola do it? The company calculates returns on its employee training costs and says some programs (Advanced Diagnostic Tools) yield $33 in cash flow for every dollar spent. Most of the programs assessed show a return in the 30 to 1 range, according to the director of training quality: for every dollar invested, there is 30 times the savings, revenue generated, or productivity gain.

Training ultimately leads to functional integration. The United Services Automobile Association (USAA) spends $19 million annually on employee training, which is 2.7 percent of its annual budget and double the industry average. One of the major decisions that CEO Robert McDermott made when he took over the company in 1969 was that "we were going to have an education and training program to upgrade employee quality and enrich the jobs." He followed through on his decision. "Today almost

30% of our people are in education and development programs—
one of the highest percentages in the country."

"We don't want anyone to get stuck in a job, especially in a job
perceived as repetitive or boring. So we do our best to get people
to sign up for training, apply for new positions, learn more jobs—
as you say, move around inside the company. Every year 45% of
our people get a promotion, and every year 50% of our people
change their jobs. We want people who know what the people in
other departments do, we want our managers to have the broadest
possible experience of the company, and we want to be able to fill
key vacant positions quickly. In other words, we want flexibility,
and we want functional integration. A company full of people
who've worked in lots of different jobs and divisions has both."

Training can turn around a troubled company. Described as "the most successful turnaround company in the history of the computer industry," Storage Technology now ships
over "99 percent" of its products on time and customers are satisfied. Prior to the company's bankruptcy in 1985, 80 percent of the
products were shipped late, and many were defective.

When I asked CEO Ryal Poppa what led to the turnaround, he
attributed it in large part to a reeducation program. After emerging
in 1987 from the earlier bankruptcy, he explained, the company
"had a ton of baggage. Even though we were in good shape,
people still remembered that we were in trouble and were almost
broken up in 1985. So I put together a team headed by a vice
president of customer satisfaction that included both maintenance
and education." They trained everybody in the company, paying
special attention to those on the front lines, especially the service
force. Employees, in turn, got the message out to customers that
"we're a quality company again."

Poppa believes corporate change is set in motion by the role
model at the top and continued by an educational process that
diffuses a set of principles throughout the organization. "We put
together what we call the operating principles focused on employees, customers, and quality." These principles were communicated through tapes, written material, and presentations by top
management to large employee groups. "Today, we're in the top
five in the industry in terms of morale."

The education function has since been combined with quality at

StorageTek. The reason, says Poppa, is that "the future is very closely tied to quality."

Training helps keep good employees. Ingersoll-Rand decided to retrain current employees rather than hire new employees with the needed skills when it modernized one of its plants from the traditional assembly line system to computerized, cellular manufacturing. "Rather than hire new, already qualified workers," explains James Sheedy, the company's manager of human resources, "we decided to retrain current employees. Most had long service with the company. They were familiar with our products, with customer complaints and compliments, and, most importantly, with the plant's long-established culture. These attributes cannot be instilled through a training seminar."

Employees responded favorably to the training program: 99 percent of those eligible volunteered to participate. In total, they spent more than 27,000 hours in the classroom, 50 percent of those hours on their own time.

"The results have exceeded expectations," says Sheedy. The education program "helped Ingersoll-Rand to profit from plant modernization." But it did so while giving veteran workers increased pride and a greater feeling of self-worth. "We also have retained dedicated, proven employees."

Training plays a lead role in assuring quality. Increasingly, training serves as the conduit to improved quality. According to *Training* magazine, three fourths of senior executives surveyed say quality is now a formal strategic goal and two thirds name training as an integral part of their efforts to achieve quality.

Like other executives I interviewed, Lou Smith, president of a division of Allied-Signal Aerospace, told me that training helps the company achieve its strategic goal to "continuously improve performance in cost, quality, and delivery." The company provides performance-based training as well as education in basic skills such as math and reading. "Anytime one can better understand what he or she is asked to do," says Smith, "then their performance is going to improve." Smith doesn't ask for tangible measures of performance improvement for the basic skills programs: "The intangibles are all I need."

The company's documented improvement in the area of quality

is impressive. According to a variety of metrics, efforts to improve quality are working: defects have been reduced nearly 2 percent, deviations are down more than 75 percent from 1986, delivery performance is nearly perfect, and supplier partnerships show a 73 percent quality improvement in the last two years.

Another example of how training impacts quality is the rigorous training program at Merck, rated the top drug company for service. Before going into the field on their own, new salespeople spend 11 months in training. They start with 10 weeks of basic medical training and three weeks learning about Merck products. For the next six months, they make field presentations under the watchful eye of a district manager. Then they spend three weeks at company headquarters refining their presentation skills. Salespeople are required to continue their education by taking medical courses at local universities.

Improving the Quality of Training

Training is currently a labor-intensive endeavor, with 80 percent of corporate training budgets going to staff salaries and overhead. The key to future improvement lies in two areas recently rated as the top two concerns of trainers surveyed by *Training* magazine: quality improvement and technological change.

I have worked with companies to assess education and information needs and revamp the training function. From my observations, there is a critical need in most companies to improve the quality of training and to adapt to rapid technological change. Here are seven basic tactics to improve training and thereby help employees learn better and avoid information overload.

1. Determine if training is needed. Training is often the assumed solution when any type of performance problem occurs. Yet "80 percent of performance problems cannot be solved by training," says Diane Graham, the training director at a large manufacturer. "The problem may be a system issue, roadblock, or poorly defined task to begin with."

A needs analysis reveals whether or not there is a problem that can be solved with a training solution and, if so, what type of training is required. Companies can husband their limited training resources by checking out the situation before jumping in with training programs. One problem, however, is that the wrong people often conduct the needs analysis; the task should be directed or conducted by instructional designers, who are specialists at this task.

A bigger issue is the lack of high-level leadership in the education and training area within most companies. Training is often relegated to at least one or two levels below vice president, unlike the corporate communication function, which often has its own vice president. This state of affairs may be a result of the view that anybody can be a trainer. In fact, the art and science of training are quite advanced today. Instructional designers with appropriate credentials are well versed in instructional science, educational psychology, learning theory, media technology, and other domains. As companies discover how crucial training is to their strategic goals, more may elevate the education and training function. By providing proper leadership in this area, companies can dramatically improve the quality of training throughout the organization, which brings us to the next point.

2. Put instructional designers, not subject matter experts, in charge of training. Companies often recruit subject matter and content specialists to provide training. The problem with that approach is that specialists are focused on the content rather than on the learner or learning process. Few content specialists know how to structure information to meet the needs of trainees. Learners often end up sitting through a lot of irrelevant information that only adds to information overload.

Instructional systems design results in more effective training. Subject matter experts provide input and may join the team, but the focal point of the training effort is the instructional designer. Designers know how to elicit the right type of thinking in learners. They're skilled in bridging that gap between what content specialists know and what learners do not know. In my role as professor, I've prepared many students earning graduate degrees in instructional design. There is a great need, however, for many more instructional designers as more companies discover they cannot afford the content-specialist-as-trainer approach.

Many large companies have adopted instructional systems design. One company that has used this approach for many years is Motorola, which provides at least 40 hours of training each year for every employee. "We have a process of instructional systems design," explains Brenda Sumberg, a manager of training at Motorola, "which up until a few years ago was quite unique. The more common model is to take a subject matter or a content expert and have that person put together the training. The problem with that," she told me, "is that content experts tend to want to tell everything they know. The participant may or may not need all of that information in order to perform the actual task," so the training adds to information overload.

3. Focus training on the learning objectives. Old training approaches bombarded people in the classroom with information that is forgotten once the person is back on the job. A main tenet of instructional systems design is to specify learning objectives or desired performance outcomes, then provide an educational experience to achieve those outcomes. Learning objectives do *not* refer to the topics to be covered during training. Rather, objectives describe the outcome behaviors that should result from the training experience. In other words, what will the trainee be able to do after the training as a result of the training experience? With this perspective in mind, training can be streamlined to achieve only the stated objectives and not burden the learner with extraneous information.

Some companies refer to this approach as "performance-based instruction," because training is focused on improving performance. The objectives are closely tied to the business need or performance issue that initially evoked the desire for training. A department seeks to improve quality by 50 percent. The sales team aims to increase sales by 6 percent. So the training objectives are, ultimately, to produce changes in the learner that can be transferred to those behaviors in the workplace.

Brief learners on the objectives. It's not enough for the trainer to know the point of the training. Research on learning indicates that knowing the objectives ahead of time facilitates learning. This basic point also applies when companies send managers to off-site seminars and educational programs. When Pennsylvania State University surveyed students in its executive development courses, less than 30 percent said they had been properly briefed before being sent to the program, but the companies said that more than 80 percent had been briefed.

4. Provide just-in-time "performance support." Another way to avoid information overload is to provide performance support that makes training available as it is needed. Ford Motor Company, for example, furnishes computers housed on rollout stands on the floor of their manufacturing line. When workers on the line have a maintenance problem, they roll the computer over, use a menu-driven screen to choose a topic, and receive training right then on how to fix the problem.

"We do not have the time or the inclination to teach everybody everything before they need it," says Ralph Adler, who is in charge of computer training and reference at J. C. Penney Co. They don't have the resources to take a "serial" approach to training—start at the beginning, give the basics, then build until everything has been covered. Performance support systems allow people working in Penney credit centers or store offices to learn just what they need, when they need it, and not waste time on anything more. The training is embedded in the company's online computer system so that learning is integrated with the task at hand.

A performance support system is timely, saves paperwork, and reduces human effort. In the past, Adler's group printed and distributed new manuals to the 1,330 Penney stores each time they enhanced the computer system's capabilities. In one year, this amounted to 8 million pages of documentation. Now they provide documentation and support online. A 2-page memo to announce an update is sent out in place of an 18-page manual. Each page saved is significant when multiplied by 1,330 stores.

In Penney's performance support system, training takes the form of online support and pop-up help windows. If needed, workers can request and receive further help from the computer system: a computer-based training course on the subject at hand. Then they can return to what they were doing at any point and avoid getting caught in an avalanche of information.

5. Involve trainers early. The best way to streamline training is to reduce the need for it in the first place. Smart companies involve trainers during the design phase of any new program or process that requires training. Trainers' input at the beginning of a project often saves many hours of work later on.

The system developers at J.C. Penney now meet with trainers before they design a new application. This is in contrast to "the way we used to do things," explains Adler, whose office provides a facilitator for the collaborative design sessions. "After they had the application all built and ready to install, trainers would be brought in and told, 'This is what we've got. How are you going to teach people to use it?' "

Collaboration at the design phase allows training to proceed in tandem with program development. An obvious benefit is a reduction in total delivery time for the product-with-training. Most importantly, collaboration results in an easier-to-use product that requires less training to master. Trainers don't waste time figuring out how to teach people to use an awkward tool. Input from trainers at the design stage expands a development group's attention to include usability issues rather than focus on "just the technical capabilities as is historically done if you let technicians do all the work."

6. Assess the impact of training. Organizations can assess the impact of training against whether or not participants mastered the targeted skills, knowledge, or behaviors. This metric for assessing the quality of training represents a significant improvement over traditional methods of simply polling learners' views. Opinion surveys ask learners if they enjoyed the training, if it met their expectations, and if they learned from it. Objective measures of learner performance offer a more valid approach than learner perceptions and opinions. The ultimate metric is change in performance back on the job as an outcome of training.

"When managers look at the kind of investment that they're making in training, they are not objecting to that investment," observes Brenda Sumberg, in charge of training quality at Motorola University. "But what they're saying is, 'Is this the right training?' and 'Is it taking?' " These questions are answered by assessing the impact of training on the basis of whether learners achieve the learning objectives.

The basic idea here is that to improve training, you need to assess how well current efforts are working.

7. Balance the practical with theory and basic skills. While it's useful to tailor training to specific performance objectives, omitting relevant "theory" is shortsighted. As the saying goes, "There is nothing more practical than a good theory." The reason is that a valid concept, idea, or basic skill may be applied by a well-educated employee in a variety of practical situations, not just one.

In the information age, white-collar workers need to be able to adapt to change and to tailor their performance to the specific situation at hand. For example, professional and technical workers are not called upon for new ideas every hour, but they need to be able to generate creative ideas when needed. It is infeasible and often impossible to train workers how to handle every possibility ahead of time. That's why training should equip employees with relevant basic skills that can be applied intelligently as needed. Some of the basic skills important to white-collar workers today include:

Reading. With the average worker spending 1.5 to 2 hours every workday reading, there is a need for better reading skills, especially the ability to synthesize information from several sources and extract meaning from written materials that are not explicit.

Math. Most workers have basic computational skills but can't apply what they know. Basic skills training is needed to help more people use math to solve problems.

Communication. Of the time spent communicating, the average worker spends 8.4 percent writing, 13.3 percent reading, 23 percent speaking, and 55 percent listening. The latter skill is often ignored in training, yet it is crucial for success in the information age.

Learning. Of all the basic skills, this is the most important, because it helps workers adapt to change. Today's worker needs to be flexible. "Central to flexibility," says ASTD researcher Anthony Carnevale, "is the ability to learn—to keep up with change, to know what needs to be learned, and to learn it without disrupting performance."

The #1 Training Need: Teaching People How to Learn

"The illiterate of the future are not those who can't read or write," says Alvin Toffler, "but those who cannot learn, unlearn, and relearn." Learning how to learn was rated as the number one basic training need by employers, according to an ASTD survey.

In my academic role, I have studied both the psychology of learning and technology of training for over 20 years. The learning techniques and tools available today are sufficient to serve your learning needs in the information age *if you use them*. The problem is a widespread lack of awareness of these solutions.

Old ways of teaching and learning—epitomized by the traditional classroom lecture—won't get you through the information age. Australian humorist Bruce Petty illustrates the old view of the passive learner. Learners start out with a *tabula rasa* (blank slate), or what Petty envisioned as an empty tub above the learner's head. The instructor is plugged into the sources of knowledge so the instructor's mental tub is full of liquids and solids representing knowledge. By opening the pipeline between the two, the learner's tub fills as the instructor's knowledge transfers to the learner.

"Well, obviously education isn't a matter of connecting pipes up and pushing things through them," said Petty. "It's something quite subtle, something 'electronic,' something like guesswork and inspiration . . . some subtle connection that no one fully understands—but every now and again it works!"

Traditional learning methods keep you thinking in the old ways—they're too abstract, passive, verbal, and artificial. Today, anything that requires rote memorization is probably not worth learning and is better left to computerization. Information-age skills are best served by highly interactive, flexible, and realistic learning environments. To use such environments, a learner must be self-directed and know how to learn.

Learning to Learn

"Knowing how to learn is the most basic of all skills because it is the key that unlocks future success," advises the American Society for Training and Development. Once you have this skill, you can achieve all of the other skills needed in the information age. Here is an overview of the three basic skills and related competencies needed to be a better learner.

1. Skill in Assessing Your Individual Differences

It pays to discover your own learning style, capabilities, and sensory preferences so you can direct your own learning efforts more effectively. The point is that every learner has a distinct or preferred way to learn. You can benefit by matching instructional methods with your preferred style and abilities. Sometimes it's better to use a method that compensates for a weakness. Take tests such as the Learning Styles Inventory, Meyers-Briggs Type Indicator, and others measures. For successful learning, you need to be able to:

- Assess your learning needs.
- Understand your own learning style and abilities.

2. Cognitive Learning Skills

Learning outcomes fall into three categories or domains: cognitive (what you think), affective (how you feel), and psychomotor (how you move). Learning to learn generally focuses on the cognitive outcomes—what you know, understand, apply, analyze, synthesize, and evaluate. Learning methods differ depending on what level of learning you need to achieve. You can "know" something by simply reading about it. However, to "apply" new information, you'll need more interaction and involvement than you get by reading. To learn effectively, you need to:

- Organize, relate, and evaluate information.
- Apply various learning strategies and tools.
- Think in both analytical-logical and holistic-divergent (big-picture) ways.
- Know how to mobilize learning resources.

3. Interpersonal Learning Skills

Although the process of learning is basically cognitive, interpersonal learning skills are important since learning involves interacting with others. "It's no good having some guy who is very wise and sits alone in a room," says *Fortune* magazine. Knowledge must be shared with others for maximum benefit. Competency in interpersonal learning means that you can:

- Give and receive feedback.
- Learn collaboratively.
- Draw from others as learning resources.

The three steps in this chapter's Action Plan correspond to these three learning skills.

Your Biggest Strengths

The following two principles and related technologies are provided to raise your awareness so you can better apply the steps to effective learning presented in the Action Plan.

Chunking and Structured Thinking

You can condense the growing glut of information into manageable units by *chunking*—combining individual items into cohesive patterns. Chunking explains why you spot license plate words— SPEEDER—faster than those made up of seven unrelated items— 1JH26P7. Both contain seven pieces of information, but you automatically chunk the letters of a word into *one* meaningful unit.

Here's another simple example: Unchunked, the following 10 numbers have no particular meaning and would be hard to remember:

3104567608

Once the numbers are chunked into three groups, a familiar pattern emerges—our phone number:

(310) 456-7608

The easiest numbers to see and remember are those that spell a word:

1-800-AKA-BOOK

Chunking is important due to limitations in "short-term" memory, the brain's capacity to retain information for a short time. Psychologist George Miller determined that short-term memory can "let in" only seven pieces of information, plus or minus two, at one time. That is why most people find it difficult to recall several phone numbers at one time unless they use a memory aid: write the numbers down, repeat them aloud until dialed, and so on.

Fortunately, there is no known restriction on how *complex* the seven units can be. A unit could be a letter, number, or simple shape. It could also be a pattern of chess pieces on a playing board, an array of blips on a radar screen, or an entire paragraph that a speed reader takes in with once glance. "It is as if we had to carry all our money around in a purse that could contain only seven coins," explains Miller. "It doesn't matter to the purse, however, whether these coins are pennies or silver dollars."

Beginners, novices, and other learners new to any area take in a penny's worth of information, whereas experts take in a gold nugget's worth. The more a person knows about a topic, the more he or she chunks information. A study of chess experts, for example, found that after one glance at a game in play, the experts were

By chunking separate items into new units, you can juggle more than before.
Source: *Design Yourself* by Hanks, Belliston, & Edwards. © 1977, 1978 Crisp
Publications, Los Altos, CA.

much better than novices at remembering where each chess piece
was on the board. It's not that the experts had superior memories
in general. When the experts viewed chess pieces arranged *at
random* on the board, they did no better than novices at recalling
locations. The experts could not rely on chunking, because there
were no meaningful patterns there to see.

The Action Plan provides specific ideas for capitalizing on your
ability to chunk information and think in a structured, top-down
way. The main idea is to discover how information on the topic
you need to learn is structured by experts and those in the know.
Then use that structure to boost your learning.

Visualization and Concreteness

The ancient Chinese told us that a picture was worth 10,000
words, and researchers are exploring just how true this proverb is
today. As you learned in earlier chapters, visual information is
easier to learn and remember. What's truly amazing is that you
actually remember visual information *better* with time, in contrast
to verbal information, which is forgotten. This visual advantage is
termed "hypermnesia."

Two psychologists discovered that people have hypermnesia for pictures when they compared people's ability to recall images (in the form of photographs), concrete words (such as "desk"), and abstract words (such as "happiness"). One group of learners saw 72 photographs, a second group saw 72 concrete words, and a third group saw 72 abstract words. The three groups were then given two chances to recall what they had seen: first on an immediate test and again a week later.

People who had seen the photos remembered most, concrete words less, and abstract words the fewest of all. The more fascinating result of the research, however, was the fact that visual memory actually improved over time, unlike verbal memory. Rather than forget, learners remembered *more* photographs after one week than they did on the immediate test.

One of the learning skills you need is the ability to select effective learning resources. One of the most important things you can do in this regard is seek out visual and concrete learning resources or create them for yourself if they're not available.

Just what are "visual and concrete learning resources," you wonder? I conducted several reviews of the impact of these learning resources and published the results in *Instructional Science* and the *Journal of Mental Imagery*. In short, you will learn more efficiently and effectively from the following:

- *Concrete words.* It's easier to learn from written texts and instructional materials that contain concrete examples and words that are easily pictured in your mind (i.e., "desk" is concrete, "form" is abstract).
- *Representational images.* Learners, regardless of learning style, are helped to some degree by representational images—those that iconically represent the new topic or idea to be learned. A direct representation can be a realistic drawing or photograph. If the idea is abstract and can't be photographed or drawn, indirect representations aid learning by showing how the idea affects something that is visible. Although this type of image is frequently used in training (when visuals are used at all), it is generally not the most effective. The following two types are more helpful.

- *Analogical images.* At the early stage of learning, an analogy helps bridge the gap between what you already know and what's new. Advertisers use analogical images to drive home their point within seconds—like showing a customer dousing the flames on his burning feet with the company's medicated foot spray. The more abstract and complex a concept, the more important the analogical image. To fully grasp a new idea, learners must move beyond limitations of analogy. Otherwise misunderstanding could result. The power of analogy lies in its ability to introduce a new idea.

Conventional programs   Object-oriented programs

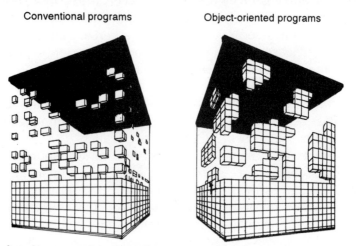

A visual analogy to explain the difference between two types of programming. "Object-oriented programming lets software developers use preassembled 'blocks' of code instead of building programs brick by brick." © 1991 by the Peed Corporation. Used with permission.

- *Abstract images.* This category includes graphs, flowcharts, Venn diagrams, maps, symbols, and other abstract drawings. Abstract images help learners extract the essence of a concept. This type of image is particularly well suited to computerization, since computers can manipulate abstract images easily including 3-D images.

Instructional Media and Technology

Today's instructional technology helps learners both chunk and visualize information. Technology for structuring information into

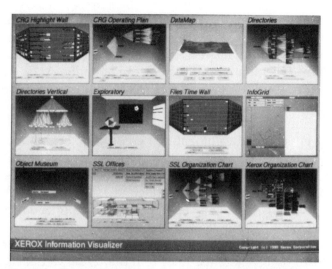

An abstract image, Xerox's futuristic Information Visualizer system provides an Overview which gives a high-level look at a collection of virtual workspaces called *rooms*. This Overview shows 11 3-D rooms and one 2-D room.

Source: Illustration courtesy of Xerox Corporation.

chunks came into popular use in the 1980s in the form of hypertext. It organizes information in a way that lets the user jump easily from topic to topic, or delve deeper for more detail on a topic. Everything can be treated as a chunk or what is termed a "node" in hypertext, from individual words to screenfuls of computer information. The chunks may integrate text with sound, graphics, computer animation, and video. This combination has led to the concept of "hypermedia," now more commonly called "multimedia."

The advantages of multimedia for learning are many. The branching capability provides shortcuts to information. One example is the multimedia program from HyperView Systems on the topic of handling hazardous materials. Selecting an explanation of how to store drums of a particular chemical brings up an image illustrating the technique. You can also pull up relevant congressional testimony at the click of a button.

A number of companies are using multimedia technology as a training or reference tool. Procter & Gamble uses multimedia to train production workers. NCR's field engineers learn how to do repairs from a multimedia program that visually illustrates the repair jobs. Federal Express trains couriers and customer service employees with interactive training technology and expects a 10 percent improvement in productivity and quality. At General Motors' Saturn plant in Tennessee, multimedia trains workers in everything from waste disposal to robotics.

Highly interactive, visual learning environments make learning not only effective, but enjoyable. I credit this technology for my understanding of statistics and hypothesis testing, two complex and abstract topics. I used interactive media to learn statistics (on the PLATO system) when I was a doctoral student at UCLA. While my classmates read statistics textbooks and memorized formulas, I interacted with dynamic computer graphics. I *saw* what the statistics meant with my own eyes. It was fun, because I controlled interactive graphics and learned from the visual feedback.

ACTION PLAN: Three Steps to Learning to Learn

The number one thing that companies can do to help individuals learn is to *model* effective learning practices during training. Many of the practices of good training are similar to the practices of good learning. Here are three steps for you to follow in becoming a better learner.

Step 1: What Is Your Learning Style?

My research with hundreds of adult learners indicates that learning is most successful when the instructional methods either capitalize on learners' strengths or compensate for their weaknesses. In several of the studies I've conducted, learners with a visual learning style or ability benefited from visual methods but not as much from an all-verbal instructional approach. Visual teaching methods also helped verbal learners but not as much as it helped visualizers.

Discover something about your own learning style by taking the following tests adapted from the *Ways of Thinking* test. There are no right or wrong answers per se. If you agree with the statement or decide that it does describe you, answer TRUE. If not, answer FALSE.

Style I	True	False
1. I enjoy doing work that requires the use of words.	____	____
2. I enjoy learning new words.	____	____
3. I spend little time attempting to increase my vocabulary.	____	____
4. I can easily think of synonyms for words.	____	____
5. I have better than average fluency in using words.	____	____

Style II

6. My daydreams are sometimes so vivid I feel as though I actually experience the scene.	____	____
7. My powers of imagination are higher than average.	____	____
8. My dreams are extremely vivid.	____	____
9. My daydreams are rather indistinct and hazy.	____	____
10. My thinking often consists of mental pictures or images.	____	____

Scoring

Items 3, 9: 1 point for each "false."
Items 1–2, 4–5, 6–8, 10: 1 point for each "true."
Add up your points separately for each style.

What Your Score Means

Style I

3 or more   You are a verbalizer. You're comfortable when you use verbal learning techniques (summarize, ask and answer questions, take notes, etc.) but also may learn when using visual techniques.

Style II

3 or more   You are a visualizer. You will learn and feel comfortable with visual resources: maps, graphs,
drawings, photos, models, charts, and concrete
examples.

Both

3 or more   You have a mixed style. You benefit from both
on both   visual and verbal learning techniques.

According to research, these two learning styles—verbalizer
and visualizer—are slightly correlated. The learning abilities they
relate to—verbal ability and visualization ability—are moderately
correlated. That is, people with high verbal ability also tend to
have somewhat higher visualization ability than the norm.

Step 2: Cognitive Strategies to Help You L•E•A•R•N

Link details to the bigger picture.   As a learner new to a
topic, it's more important for you to find out how the field is
organized than to memorize facts. For years, I've watched learners
suffer from information overload when they try to master a hodge-
podge of facts. By linking details to the bigger picture, you draw
on your ability to chunk information.

- Determine the organizing structure of a new field *first*, then
  use it to guide subsequent study.
- Thoroughly learn the structure before you learn the facts
  just as builders make a solid frame for a house before they
  nail on the siding. Try it the other way, and the facts won't
  stick, because there is no underlying framework to hold
  them.
- Find out how experts in the area structure the information.
  Discover how they see the big picture by looking for
  taxonomies and categorization schemes.
- Take notes as you read, listen to lectures, or work through
  instructional media. Organize and relate your notes to the
  important categories.

- Outline information as you learn it. Consider adding an outlining tool to your personal computer as an instantly available desk accessory.
- Relate details to some higher-level concept. List them in your notes by bulleting two- or three-word descriptors of each.
- Draw maps and charts to illustrate how main ideas relate to one another.
- Think about the learning objectives throughout the learning process. Ask yourself, "Why is this material important? What will I be able to do as a result of learning this?"

Enliven learning. Confucius said, "I see and I forget. I hear and I remember. I do and I understand." Today we say that people "learn by doing." Yet passively watching a video and listening to a lecture remain the two training methods used most often today in U.S. companies of more than 100 employees. Become an activated learner and dramatically improve your learning.

- Ask yourself questions about the learning material and answer them (it's OK in this situation). You also can ask your trainer or others who know.
- Organize your learning efforts around a sample test or study test. If none is available, make up your own.
- Never sit down to read and study material from start to finish without stopping. You will learn how to be an active, top-down reader in the next chapter.
- Restate verbal information in your own words by summarizing, paraphrasing, and making analogies.
- Translate abstract ideas to concrete terms. Think of specific examples that illustrate the new idea.
- Apply new information as soon as you can in a setting that gives feedback on how you're doing. Motorola provides facilitators who give feedback to managers as they apply what they've learned about how to implement change in an organization. Based on the feedback, managers can learn from their mistakes and adjust their efforts.
- Take brief notes to improve your active listening skills during lectures and tapes. Go back and review the notes, filling in the gaps as needed.

- Visualize ideas mentally, especially if two ideas must be associated.
- Simulate or role-play real-life situations to improve your ability to transfer skills learned in training to the workplace where the skills will be applied.

Analogize new ideas to something familiar.   New ideas will be clearer and easier to understand if you can find or create relevant analogies. Building an analogy bridge allows a smooth transition to the new information, rather than falling into the murky waters of confusion. As Goethe pointed out many years ago, "Everyone hears but what he understands."

- Ask trainers to make analogies when they present any new concept. Since the trainer already understands the new concept, this task is much easier for him or her to do.
- Literally translate the words you are reading into a visual image that the words imply. For example, the visual analogy on page 176 translated the terms "blocks" and "build" into a literal picture of a brick wall, built with individual bricks (for conventional programs) or blocks of bricks (for object-oriented programs).
- Match attributes or characteristics of a new concept with a familiar concept that also has some of these attributes. The visual analogy of the juggler on page 174 shares the common characteristic (with the chunking concept) that separate items are grouped together into one unit.
- Distort or invert attributes of a concept. The "hydraulic" analogy for education on page 170 is really making the opposite point to the one the picture conveys. The cartoonist is saying that education is not simple and mechanistic, but rather an intuitive process that is difficult to define.
- Look for cartoons associated with key points you are trying to remember. The reason is that cartoons help you connect "new ideas in the cartoon to old ideas" from your experience, according to researchers.
- Ask trainers to illustrate their messages with cartoons. Researchers say that cartoons are a very efficient means of giving an example. They can convey a point in a matter of seconds in a way that makes a lasting impression.

Reduce information to a simple representation.   Albert Einstein advised, "Make everything as simple as possible, but not more so." He also believed that if you can't explain something simply, then you don't really understand it. Productivity in the workplace depends on being able to simplify, not complicate, information.

- Create an abstract diagram to map the relationships between key elements. This could be a Venn diagram, a flowchart, or organization map that serves to simplify complexity.
- Make sure your visuals are relevant and appropriate. Visuals can confuse or clarify depending on how you create them.
- Extract the key ideas from information and use them to organize the details.
- Use formating cues in text to pick out main ideas at a glance. In this book, for example, boldface type and layout distinguish headings and subheads while details are found in bulleted lists—like this one.

Nest learning in a hierarchy of challenge.   The benefit of nesting is twofold: it locks the information into your knowledge structure so it won't get lost (forgotten), and it motivates you to keep learning. Take a look at any television game show, and you will see all of the elements of hierarchical nesting that keep contestants challenged and viewers captivated. (I know because in my days as a graduate student, I was a contestant on *The Price Is Right.*)

- Set multiple-level goals. The thing that keeps a person motivated is to strive to get to the next section or level. (When I first heard those words, "Come on down," I was motivated to win the first game and make it up to the stage—which I did.)
- Tackle increasingly difficult questions and problems as you continue learning. As you learn more about a new topic, it gets easier—and you're more apt to get bored. Keep yourself challenged by attempting the more difficult tasks as you go along. (Once I made it up on stage, the questions got harder. Luckily I could look at my family in the audience for the right answers—which I got.)

- Keep an open mind to serendipity and chance as you learn. Learning is not a matter of memorizing cut-and-dried facts. If we knew everything that was going to happen, learning would be boring. It's the unknown element of chance that keeps things exciting. (As I pulled hard to spin the wheel, I waited excitedly to see if my spin would win—which it did.)
- Hide some information and make yourself guess or "work" to see it. It's more challenging when you don't know everything up front. Hidden information makes you use your curiosity to motivate further study. (In the "grand showcase," I was motivated to find out the prices of items so my bid would win—which, alas, it did not.)

Step 3: Learn Collaboratively

More than ever, work is a group activity as white-collar workers pool their talents to pursue common goals. The real value of group work lies in the fact that it is more than the sum of the individual efforts. Peter Senge, author of *The Fifth Discipline*, believes that team learning is a tool for raising the collective IQ of a group above that of anyone in it. He says team learning is vital, because "teams, not individuals, are the fundamental learning unit in modern organizations; unless the team can learn, the organization cannot learn."

There is an interaction—a synergy—that results. I experienced such a synergy in the 1970s as I collaborated with three other doctoral students at UCLA to study for our comprehensive exams. The exams were notoriously difficult, because they required doctoral students to have a broad understanding of the field before starting dissertation research.

- Assess the needs and capabilities of other team members and determine how each can contribute to the team effort. In the UCLA study group, we quickly determined who knew what and put each person in charge of the areas he or she knew best.
- Establish goals as a team. Communicate these goals to everyone on the team. The four members of our study group agreed that the "expert" in each area would be

responsible for giving the group a handle on that area. To do this, the expert prepared a 2–3 page summary and led a discussion of key issues, research, and conclusions in the area.

- Discuss ideas, ask questions, provide feedback to others, and ask them to give you feedback. Our study group sessions were lively to say the least. "I don't see why that's so important." "Where in the world did you come up with that answer?" "You're giving us too much detail. Get to the point." "Get rid of these wordy paragraphs. Lists tell me all I need to know."

- Check out the many ideas for group collaboration in the other chapters of this book.

By the way, everyone in our study group passed on the first try—an amazing feat when many of our fellow students had to retake portions of the exam and some failed outright.

SELF-TEST: What Kind of Learner Are You?

Take the test on page 179 to find out your preferred learning style.

Coming up. The average American worker spends 1.5 to 2 hours each workday reading. Although 92 percent of Americans said in a Gallup poll that they think reading is very important, most of us would like to spend less time reading and writing and get more out of the time we do spend. To find out how to extract the essence, turn to the next chapter.

Chapter Seven

Too Much to Read & Write? . . .

Extract the Essence
with Top-Down Techniques

I am not a speed reader. I am a speed understander.

Isaac Asimov

The words of my book nothing, the drift of it everything.

Walt Whitman

You most likely have piles of correspondence, trade magazines, newsletters, journals, newspapers, and other material in your office right now waiting to be read. Add to that your boss walking through the door to tell you to produce a report by next week that highlights opportunities in the waste treatment industry since the company plans to diversify into that area. You need to read about potential acquisition candidates and do a comparative analysis of the expenditures, market shares, and future trends within this industry. If you haven't maximized your reading and writing strategies for the age of information, you'll soon slide further into the danger zone of information overload.

This chapter shows you how to drastically cut your reading load, boost your reading speed to meet special demands, and write in a way that saves time for yourself and for the reader.

Cut Your Reading Time . . . with a Clipping Service

A clipping service is one of the best-kept secrets of many top leaders, CEOs, and other executives for reducing their reading time. The typical clipping service employs hundreds of readers to peruse more than 17,000 publications and send you clippings several times a week on any topics you choose. It's like hiring hundreds of pairs of eyes to do your reading, and the cost is probably less than you'd expect. The base rate starts at about two hundred dollars a month, plus a small charge for each clipping that is sent. If you think about what your time is worth, the service generally pays for itself easily.

Here's how it works: You provide the service with key words to guide the readers. A key word can be a company or person's name, a topic of interest, products, competitors, or any other subject. Your key word profile is entered into a computerized databank, and hundreds of readers refer to the profile to find relevant clippings. They'll even read specialized trade and professional publications that you need to follow. Articles are literally clipped out of the publication (or copied if other clients want the same information), tagged with the date and source, and quickly mailed (or faxed) to you.

Scene of a reading room of a clipping service.
Source: Courtesy of Burrelle's Information Services.

In the past, clipping services were used primarily by public relations to track publication of their press releases. Now nearly a third of those who use a clipping service do so to get information in a particular subject area.

"The clipping service keeps me current in my specialized topic," Roger Herman, CEO of a multistate training company, told me. He spends about $4,000 a year on the service and says it's a good investment. "I'm on the leading edge of what's happening. In my business, I have to stay abreast of everything that's happening in my field." Herman's company uses information from the clippings in a variety of products and services including a regular newsletter for clients, seminars, public speaking, books, and other training materials. He has authored four books (including *Keeping Good People*) and 350 articles, and still finds time to read 30 periodicals and four or five books per month.

People employ a clipping service not only to save time but also to leverage their labor. Here are just a few examples of how executives and others are using a clipping service. You can:

• Receive the late-breaking news each morning by fax about topics you choose with services such as *NewsExpress* from Burrelle's Information Services. The stories come complete with photos. Executives are using this service to stay up with press coverage of their own company as well as their competitors, current developments in their specialty, and other late-breaking news pertinent to their business.

• Get a jump on printed news by receiving it directly from the major wire services before it is printed or broadcast. The clipping service reads the newswire services—Dow Jones, Reuters, Associated Press, United Press—and sends you the complete, unedited, original story before it's altered by the biases or space limitations of a publisher. Some companies also monitor foreign-related newswires such as the Japan Economic Newswire.

• Find out how much your competitors are spending on advertising and in what markets.

• Receive summaries of articles from trade journals or consumer magazines. Of course, the service also provides you with a clipping of the full article.

• Follow foreign markets and publications. The monthly fee is modest, generally $10 per month per country.

• Read transcripts or view videotapes of television news or other information on broadcast or cable TV. Most clipping services now monitor television news and can tell you how many times your company or topic of interest was mentioned and on what program. Companies use this service to keep track of news coverage of their products or competitors' products.

Video "reading room" where specialists enter programs into an online database available one day after broadcast.
Source: Courtesy of Burrelle's Information Services.

Check local listings in your business-to-business telephone directory to contact a clipping service. Major national services are listed in the resource directory.

Alternatives to a Clipping Service

The main benefit of a clipping service is that you receive news and information tailored to your specific needs by simply picking up the phone and placing your information order. Check out the following ideas to keep up with the latest business information in a way that can be faster, more tailored, or less expensive:

• *An audiocassette news service.* A business news service on audiocassette called *Newstrack* culls newsy items from 150 business publications at an annual fee of $250—less than one tenth the cost of a clipping service. The limitation is that the news is not individualized to your needs.

• *An online information search service.* You can receive up-to-the-minute news from the newswires on your personal computer with *First Release* from Dialog Information Services or *Executive News Service* from CompuServe. You can receive information on any topic once it becomes available with *Current Awareness Service.* The purpose is to keep you updated on your topic of interest.

• *Connecting directly with the collective wisdom.* Why spend hours reading something that others could tell you about in five minutes. Ask those in the know and eliminate the need to read the information for yourself. Or, at least, let others guide you to the best written source. If you have a specific question, try posting it on an electronic bulletin board or forum. Ask an expert. Tap your network. Then spend your time with those materials that you *must* read yourself.

The information explosion is forcing some groups to abandon traditional reading materials such as magazines and journals in favor of direct, instant communications. Scientists, for example, are turning to electronic networks to stay current. Physicists working in certain fields are most affected as new research results are being produced weekly or even daily. Printed journals are out of date before they are printed.

There is a danger, however, in relying on electronic media. "The explosive growth in this type of [instant] interchange," cautions a task force of the Libraries Research Group, "implies a more casual approach to the acquisition of information and may ultimately pose a threat to the orderly reporting and maintenance of the records of research results." The lag time in printed information is due in part to the many controls in place to assure high quality. Electronic and informal communications are slowed by no such controls, and therein lies the danger.

Tips to Reduce Reading Load

It's *not* a good idea to cut back indiscriminately on what you read. The reason is that reading can *save* you time, because it gives you the opportunity to learn from other people's experience. The great philosopher Socrates said, "Employ your time in improving yourself by other men's writings so that you shall come easily by what others have labored hard for." A Gallup poll once found that the average person listed in *Who's Who* reads 20 books a year. Yet 58 percent of Americans never read a nonfiction book once they get out of school. The point is that reading is good for you, as 92 percent of Americans would agree.

The problem, of course, is that too much of a good thing can be harmful. You can drown trying to keep up with a sea of reading. The solution is to reduce your reading load yet still get the information you need by extracting the essence.

"Surviving and thriving as a professional today demands two new approaches to the written word," advise Jimmy Calano and Jeff Salzman of CareerTrack. "First, it requires a new approach to *orchestrating* information, by skillfully choosing what to read and what to ignore. Second, it requires a new approach to *integrating* information, by reading faster and with greater comprehension." You can begin to orchestrate what you read with the following three steps to eliminate nonessential reading.

1. Pare Your Subscriptions to the Essentials

"I receive an incredible amount of mail and a lot of it is valuable information," marketing director Miles Babcock of Time/Design Corporation told me recently. "I don't want to toss it in the trash without looking at it thoroughly." It could mean throwing away an opportunity. To get the most from his subscriptions with the least time investment, Miles subscribes to a few key industry publications, scans and marks interesting material, then later reads what's marked and pitches the rest.

Review your current subscriptions and eliminate as many as possible. Ask yourself these questions to evaluate which ones to eliminate:

- Does this publication contribute to the big picture? Will it help move me toward my main goals?
- Could I replace several trade journals with one "industry watchdog" publication that provides an overview and commentary on the industry?
- What would happen if I discontinued reading this?
- What subscriptions offer the most value for the least amount of reading time?
- Could I refer to this publication when needed in my organization's library?
- For company subscriptions, would I be willing to pay for this subscription myself?
- Are my colleagues already reading this, and could we share reading responsibilities?
- Does a month pass before I look at this publication after it arrives?
- Could I get the same information through a more efficient means—online search, forum, or collegial network?

Based on your answers, select only the top few and forget the rest or you'll risk information overload.

2. Avoid the "Urgency" Trap of Newspapers

Newspapers pile up on a daily basis, adding to the clutter and urgently demanding your attention for information that may not be important. Before you invest your time reading a newspaper, ask yourself, "Why am I reading this paper?" If your purpose is to keep up with important news, realize that the typical daily contains only 13 percent "hard news" and 60 percent ads. Consider the alternatives.

Reason for reading	*Better way*
• Keep up with late-breaking headlines	• Headline news on cable TV • Faxed headlines from clipping service: *AM Newsbreak, NewsExpress*

	• Online newswire database retrieval: *First Release, Executive News Service*
• Follow late-breaking business or financial news	• All news station on cable TV: *CNN*
	• Financial and market coverage on TV
	• All news radio
	• Faxed news from clipping service
	• Online database retrieval: *Current Awareness Service*
• Monitor business news	• Weekly business magazine
	• Business news summary on audio: *Newstrack Executive Tape Service*
	• Clippings twice a week from service: Luce Press Clippings, Burrelle's
	• Online search: Dialog, CompServe, BRS
• Learn helpful business tips	• Book summaries and digests: *Soundview Executive Book Summaries*
	• How-to audio cassettes: CareerTrack; Nightingale-Conant
	• Join an online user group or forum

Today, you can access most newspapers online. That allows you to set up your computer to pull the type of information you want automatically from the paper. You can print out the information that your computer retrieves in this way so you get the best of both worlds—the accessibility of electronic media and the legibility of paper. The catch is that it will cost you several hundred dollars per connect hour to retrieve information from newspaper databases, and many of them do not have up-to-the-minute news. The news may be as much as three days old.

Are there *any* legitimate reasons to read a printed newspaper (other than to check the ads)? Yes, there are three main reasons. For one thing, you could read a paper to check details or see illustrations of stories abstracted in other media. *The Wall Street Journal* or *The New York Times* are appropriate sources for checking details of information abstracted elsewhere.

The second reason to read a newspaper is to gather unexpected but useful information. Online services may be more efficient, but they require you to specify exactly what you want up front. Yet, most managers can't predict exactly what information they'll need, according to research conducted at the University of Minnesota. The problem is illustrated by the difference between looking up a number in the phone book and calling directory assistance. For directory assistance to retrieve it electronically, you must know the party's name and spell it correctly. For illustrations and information on unexpected topics, *USA Today* is my favorite, because the information is structured and accessible. Main points are often bulleted, and there are plenty of maps and illustrations to put the information in context.

The final reason to read a paper is to find information not available elsewhere. For example, it makes sense to read your neighborhood newspaper to stay current with what's happening in your local community.

3. Streamline All Other Reading

Transfer as much of your reading load as possible to others, so you can concentrate on essentials. There are a number of things you can do to minimize your reading time yet maximize your reading return.

• *Delegate part of the reading task to others*. This strategy won't work for everything, but some material can be screened or preread by others. Have your secretary or assistant scan, then post an action note on each item that needs additional reading by you.

• *Share reading responsibilities with others*. Some managers use teamwork to cut back on the amount of reading each individual must do. One such executive at Mellon Bank couldn't keep up with all the reading his job required, so he asked his staff to read important books and summarize them in four or five pages. Staff members read titles like *Competitive Advantage of Nations & Their Firms* and *Beyond the Deal: Optimizing Merger and Acquisition Value.*

"I couldn't possibly read all these books," the manager admitted. "But if the books were interesting enough, we'd present [them] at staff meetings."

• *Request that correspondence begin with a statement of purpose.* Have others tell you up front what they want you to do, especially for internal correspondence and reports.

• *Encourage color coding in your organization*—pale blue paper for printing the departmental meeting notices and minutes, red for "hot" news or trouble, light green for informational (FYI) memos. I color-code seminar handouts for my university classes, according to rainbow colors: handouts for the first week in red, orange for the second week, yellow for the third, etc. Nearly all of my students are working adults who must miss an occasional class because of work commitments. With color coding, they can easily keep track of what they missed.

• *Ask writers to include a one-page executive summary of a long report or other document.* Or give them specific questions you would like answered. This approach is not foolproof, as one city councilwoman discovered. She sent a list of questions to an agency, asking them to clarify their budget report. "The response totaled 150 pages, twice the size of its budget." If you must read a lengthy report that has no executive summary, read the opening and ending sections. They generally list the purpose and conclusions.

• *Read summaries, reviews, and digests that condense business books and articles.* Subscribers to one service, *Executive Book Summaries* from Soundview, receive two or three 8-page book summaries every month. The company's editorial board selects the top business books and summarizes their key points. According to Soundview, each summary takes about 15 minutes to read and is "a skillful distillation that preserves the content and spirit of the entire book" unlike reviews (based on opinions) and digests (book excerpts strung together). Check the resource directory for companies that offer summaries, reviews, and digests.

• *Use audiotapes to leverage your reading time,* especially if you commute. You can find everything from an audio summary of current business news to an enormous library of how-to cassettes. A find for those who want to stay current but have limited time and money is the biweekly *Newstrack Executive Tape Service.* The service monitors about 150 publications including *Business Week, Forbes,* and *The Wall Street Journal* among others and summarizes top stories in two 90-minute tapes per month. The thing I like about the service is that you can also order printed copies of the articles included on the tapes. Catalogs of how-to cassettes in business are available from CareerTrack and Nightingale-Conant. Check the resource directory for more information.

The Easiest Way to Read Faster

You won't solve your reading dilemma by simply cutting back on your reading load. You must boost your reading speed to remain competitive. Reading at the average rate of 250 words per minute, it would take you nearly a day to read just one 300-page book. Fortunately, you can apply big-picture thinking to the reading process.

The number one way to improve reading: Apply techniques of *top-down reading* to bring the power of big-picture thinking to the reading process.

Top-down reading is a matter of maintaining your focus on your own big picture, then spending just enough time with the reading material to get what you need. Michael McCarthy, author of *Mastering the Information Age,* says, "For most of your reading, the essence—not every detail and every word—is all you really need." Your reading speed increases dramatically when you learn how to pick out the essence rather than read word for word through the material. Once you learn how to get "the essence of a 100,000-word book in ten minutes," says McCarthy, "your effective reading speed would be 10,000 words per minute—not bad!"

Technology is making it easier to get right to the point. Computer-based reports using hypertext or hypermedia allow readers to jump right to needed information.

• Top management at GTE headquarters in Stamford, Connecticut, switched from their 200-plus-page paper reports to a hypertext setup. Now the executives access just the data they're interested in. Soon after the conversion to hypertext, the president insisted on getting all reports in this format.

• Domino's Pizza produced an interactive, animated version of its annual report. Readers select the area of interest, then watch trucks drive across the screen to dramatize sales increases or listen to quotations being narrated. The hypermedia report not only speeds "reading" but also makes it more fun.

• Many computer-literate managers streamline reading by directing the computer to retrieve needed information and store it into appropriate hypertext nodes or "in-baskets." By linking hypertext capabilities to online information retrieval services (discussed in Chapter 4), you can create your own reports, such as news reports, that are faster to read because you can pinpoint just what you want quickly.

The following Action Plan explains the five easy steps of top-down reading. In most cases, this method is all you really need and allows you to skip word-for-word reading altogether.

ACTION PLAN: Five Steps to Top-Down Reading

The error that many readers make is to jump right into reading an article or book at the beginning and read every word until they come to the end. Reading specialists have long taught a five-step method for better reading called SQ3R: Survey, Question, Read, Recite, and Review. Current approaches build on this basic method to preview before reading.

The key to better reading is to be a productive rather than a passive reader. You'll get more out of what you read if you literally *produce* meaningful connections between what you already know and what you're reading.

The following top-down reading method will help you read faster and remember the content longer. It is based on the proven importance of top-down thinking in cognition. The Top-Down Reading system consists of five basic steps: Rate, Question, Survey, Skim, and Summarize (RQ3S).

1. *Rate the Value and Reasons for Reading*

Each time you pick up a report, article, book, or letter, rate how important your reasons are for reading it based on the title or the source *before* you read any further.

Rating helps you focus on the big picture and judge how much of your resources to devote to reading. Rate your reasons for reading both before and during reading. Ask yourself:

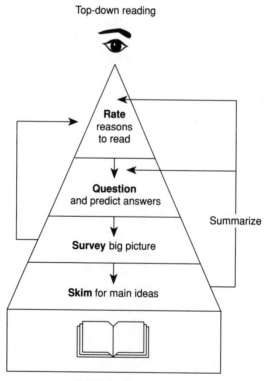

© 1991 by K. Alesandrini.

- Why should I read this?
- What do I want to understand or accomplish in relation to the big picture from reading this?
- Will reading contribute to my big-picture goals?
- Can I compress my reading time if I first delegate the initial reading task to an assistant, clipping service, or other information provider and then read what they have extracted—that is, read only the essence that has been extracted?

2. *Question Yourself and Predict the Answers*

The second step is to take a few minutes to pose specific questions that you want answered by the written material. This keeps you actively involved in the reading process. Once you begin reading, you won't simply rely on the questions and issues raised by the author. You'll have your own specific reasons for reading—to get answers to *your* questions.

Once you've formulated questions, anticipate the answers based on what you already know. This is important, because asking questions is fairly easy, but trying to answer them means you'll really have to think! It doesn't matter if your answers turn out to be wrong. The value of this activity lies in the type of thinking it triggers, not in the correctness of your answers.

Suppose you came across an article entitled, "A Taxing Matter: When Is a Worker an Independent Contractor?" (This actual article appeared in the *Journal of Accountancy*.) If reading this article can further your big-picture goals, then begin to formulate questions: "Is an independent contractor defined by number of hours worked per week for a company, by location where the work is performed, by who furnishes the tools? What penalties does the IRS levy for an incorrect classification?" Make your best guess at the answers: "A worker cannot be classified as an independent contractor if he or she works more than 20 hours a week at our company site using our equipment. Penalties may run as high as 25 percent." These answers are not exactly correct, but they get you thinking along the right lines, arouse your curiosity, and give you something to compare the correct answers against.

3. Survey the Written Material

At this point, you should spend a minute or two to survey the material to apprehend the essence. The point of surveying is to get a sense of the whole. Find out what the general theme of the article or book is and essentially how the writer has organized it. As you survey, you can determine how much effort it will take you to benefit from it. Reading will be slow if the topic is complex and unfamiliar.

You should rerate the value of reading this material once you have surveyed it. After you have a feel for how long it will take to get the answers you seek, ask yourself if the time required is worth it in light of your big picture. You may also need to reformulate your questions after surveying the material. The original questions may have been too broad or too specific and need to be modified to better fit the reading material. To continue the previous example, you would add the question, "What can I do to make sure our company doesn't encounter this problem?" Then predict the answer, "We should have our contractors sign a properly worded agreement." Once you have surveyed the material and formulated questions and answers, you'll be able to get answers to your questions in most cases just by skimming rather than reading word for word.

Surveying an article should take only a few minutes, a book 5 to 10 minutes. Here's how to do it quickly:
• *Read the book jacket and cover blurbs of a book.* This is the best way to get the point of a book and find out something about the author. Realize, however, that recommendations and claims on the cover of the book are marketing appeals to sell the book.
• *Begin by reading the summary or abstract* of an article. If it contains neither, read the first and last paragraphs. They will give you an idea of the writing style and often state the main points and conclusion.
• *Assess the author's qualifications.* Skim the description of the author, usually found at the back of the book or end of the article. Is the author qualified to write on this subject? Is a new or un-

known author endorsed by a reputable publisher or by someone with credibility? Remember that misinformation can be worse than no information. You may want to skip reading the material if credibility cannot be established.

• *Look over the table of contents* of a book or long report. This tells you how the material is organized, how many sections or chapters it contains, and what the main topics are. Think of the table of contents as the big picture.

• *Check the references, bibliography, and index.* You can get a pretty good idea what topics are covered by flipping through the index. The bibliography or chapter notes tell you what sources the author used in writing. Articles with no references are generally written in a casual style that may reflect a casual attitude toward communicating the facts. Conscientious authors give their sources.

• *Flip through the material briefly.* Survey the content by noting illustrations and major headings. If the author has visualized and highlighted the main points, you'll be able to pick up the essence very quickly.

4. Skim the Written Material

Skimming is similar to surveying except on a smaller scale. To skim, look for the essence of what each subsection or paragraph conveys and skip less important material. Keep in mind the specific questions you want answered. The goal is to read for answers to those questions. If a chapter or section does not address those questions, skip ahead. "He has only half learned the art of reading," said former British prime minister Lord Balfour, "who has not added to it the more refined art of skipping and skimming."

• *Skip to relevant section headings* that can answer the questions you posed at the outset. Skipping what's irrelevant boosts your reading speed more than anything.

• *Read the first sentence* or two of any section that seems relevant from the heading. If it is indeed relevant, then read the first sentence of each paragraph in the section. Since the first sentence is usually the topic sentence, you can assess whether to read more sentences in that paragraph or skip to the next paragraph.

• *Glance at the last sentence* in relevant paragraphs to see if it gives an important conclusion or wrap-up.
• *Don't read supporting material* found within paragraphs unless it answers your questions.
• *Intersperse skimming with summarizing* as you read. After you skim a relevant section that answers one of your questions, stop to summarize as explained in the next point, then continue to skim.

5. Summarize as You Skim

Merlin Wittrock, professor of educational psychology at UCLA, recommends that readers generate associations between what they read and what they already know. He and I conducted a study of how to improve reading comprehension among adults. Readers in our study remembered what they read when they had to summarize it in their own words. The study also tested the usefulness of asking readers to generate analogies to the new information. Most readers found it more difficult to create an analogy than to summarize, but both techniques helped reading comprehension.

• *Verbally restate or write notes* of the main points *in your own words.* The important thing here is to avoid using the terminology you just read. Translate the information to simple terms or familiar ideas.
• *Visualize or sketch the main points.* Abstract graphics (charts, diagrams, graphs, maps, symbols, etc.) that you learned about in the previous chapter are useful for summarizing what you've read.
• *Answer your initial questions* as you skim the material. Again, summarize the answers in your own words rather than take the author's answers verbatim.

If you apply this five-step process, there should be no need to read word for word. You'll have the answers you need and avoid wasting time on material that does not fit into your big picture.

Writing at Work

Written communication plays an important role in the workplace, although only 8 percent of the average worker's communication time is spent writing. "Written communications become part of a relatively permanent information base," cautions ASTD researcher Anthony Carnevale. "Inaccurate or unclear writing can pollute the shared information base and affect the quality and efficiency of work upstream and downstream." It's important for white-collar workers in the information age to learn to write with brevity and clarity.

The Trouble with Corporate Writing

Do you agree with the majority of managers who say that they could do without many of the reports they receive, but consider the ones they write to be essential? According to one estimate, only 20 percent of managers consider their own reports to be necessary but complain about the ones they must read from others.

Many top executives and several recent United States presidents have refused to read any communication longer than one page. Some writing experts discourage verbosity and urge business writers to be brief. But the biggest problem with corporate communications is not excessive wordiness, say John Fielden and Ronald Dulek, authors of *Bottom-Line Business Writing*. They contend that length matters little as long as readers can understand the point—the bottom line—quickly.

"As a result of an in-depth study of the writing done at the division headquarters of a very large and successful company," explain the experts, "we are absolutely convinced that advice such as 'Be brief!' is not only useless, it does not even address itself to the real writing problem.

"What causes trouble in corporate writing is not the length of communications (for most business letters, memos, and reports are short), but a lack of efficiency in the organizational patterns used in these communications. Put simply, people organize messages backwards, putting the real purpose last. But people read frontwards and need to know the writer's purpose immediately."

The main error writers make: they don't start by telling their purpose and what they expect of the reader.

Only 5 percent of corporate communications are organized properly, according to the evidence. Fielden and Dulek examined

2,000 letters, memos, and reports randomly drawn from company files and found that 95 percent had to be read and reread to be understood. Wordiness was not the problem. The typical communication was only one page long. It was rare to find a report longer than three pages. Most of the documents were well written in terms of mechanical correctness and appropriateness of word choice. However, the documents did not communicate the essence clearly. The reader had to ferret it out.

Redefining "Brevity"

Surely the best thing you can do to speed reading and writing in your organization is to redefine brevity in terms of the time it takes the reader to *extract the essence.* "Brief" memos and reports, then, are those that are quickly understood, not necessarily those of few words. And the best way to write something that is quickly understood is to begin with the purpose or main point, because it allows the reader to form a big picture that speeds reading.

Try the following activity yourself, then use it to persuade others in your organization that short memos are not always the easiest to read. The first memo is based on an actual memo sent in a large corporation. Keep track of how long it takes to read and understand each memo.

Memo A

The first of a series of meetings of the Strategic Marketing planning group will be held on Thursday, September 7, from 1 to 4 P.M. in Conference Room C. The purpose of these meetings is monitoring and suggesting changes in overall market strategies and product support. Attached is a list of those managers who should attend regularly. They should specifically be prepared to review alternative strategies for the new product line. The purpose of this reminder is to ask your help in encouraging attendance and direct participation by your representatives. They should contact Frank Persons for any further information and to confirm attendance.

Memo B

Please encourage your managers whose names are listed on the attachment to attend and participate in the meetings of the Strategic Marketing Planning group.

The next meeting is to be held on Thursday, September 7, from 1 to 4 P.M. in Conference Room C. Please have your representative(s):

1. Contact Frank Persons for any further information and to confirm attendance.
2. Be prepared specifically to review alternative strategies for the new product line.
3. Be ready to discuss changes in overall market strategies and product support.

Both memos are under 100 words, but the second memo is obviously easier to read. Why? It not only starts with the purpose but also lists how to respond. Memo A buries the real purpose and requested action in a rambling paragraph.

Big-Picture Writing

Many business writers today think that technology can make them better writers. Computers check spelling automatically, provide lists of synonyms at the touch of a key, and format documents automatically. But using a computer can make it *more* difficult for you to see the big picture. "The computer has been endowed with

© 1991 Brad Veley

features that diminish the quality of writing," claims *The New Republic*, "causing writers to focus on lines and sentences and lose sight of their work as a whole. On paper you can glance at many pages at once. On a computer screen, you are confined to a matter of lines." Technology is no easy answer to faster, better writing; seeing the big picture is.

Computers and other technological aids to writing cannot substitute for your own big-picture thinking. Technological tools do speed the writing process along to be sure. But technology lacks the good judgment and top-down thinking needed for effective writing.

Writing with a Smile and a Simile

If a picture is worth a thousand words, picturesque language—analogies, metaphors, and similies—are worth at least a few hundred. Used creatively, these devices will help you get your point across with punch and have fun doing it.

A client of ours we'll call Joan, who worked at a large oil company, used figurative language to make a point without getting caught up in a lengthy debate. Joan was responsible for a software development project but had problems getting a key manager we'll call John to commit the necessary human resources to her project. The purpose of the project was to create software to answer tax questions for a group of analysts in the company. To document several problem areas, Joan sent John a memo outlining the need to involve the analysts who would be the eventual users of the software—the "end users"—during the design phase of the project.

John disagreed and saw no need to involve the end users (the intended "customers" of the software) during the design phase of the project. He wrote a three-page memo to say so.

Joan laid the matter to rest by writing back: "As the project team that gave the world the Edsel discovered, it is dangerous to ignore the end user in the early phase of project development." Her strategy to link the software development problem to a well-known fatal flaw (no consumer input on the Ford Edsel) worked. She not only won the involvement of the analysts during the design phase, she headed off what was turning into a memo war with John. Figurative language packs a gentle, but firm, wallop.

ACTION PLAN: **Five Steps to Top-Down Writing**

Writers often make the same error that many readers make: they sit down and write word for word from beginning to end. This approach often results in rambling documents and frustrated writers, not to mention confused readers. The problem occurs when a writer strings together a lot of details without a coherent view of how they fit together into the big picture.

The principles of top-down reading can be adapted to the writing process. Top-down writing means that you first tell the reader the point or overall topic and then fill in the details. This approach encourages top-down processing by the reader, a type of mental thinking that keeps the reader organized and able to remember details. Here's how to adapt the RQ3S method to the writing process.

1. Rate the Value and Reasons for Writing

Rating makes you focus on the big picture. If the reasons for writing are trivial, don't waste your time. Rate your reasons both before and during writing.

• Make a mental note of your reasons for writing a memo or other correspondence, but don't feel you must tell the reader your reasons as *you* see them.

• Tell the reader why this material has been written *from the reader's point of view.* Explain why the reader should find it valuable.

• Decide what it is you want the reader to do after reading what you've written. What will the reader know after reading a letter or report? What action do you expect?

• For long reports, begin by writing an executive summary that gives the major conclusions in a page or so.

• For shorter memos and letters, begin by stating the purpose or goals, or you could make your point with an analogy.

2. Question Yourself and Give the Answers

The second step is to pose specific questions that you want to answer by writing. This keeps you focused with specific reasons for writing—to answer the key questions.

Once you've formulated questions, communicate these questions or issues to the reader along with the answers.
• List key questions that you're addressing so the reader has a clear understanding of the main points.

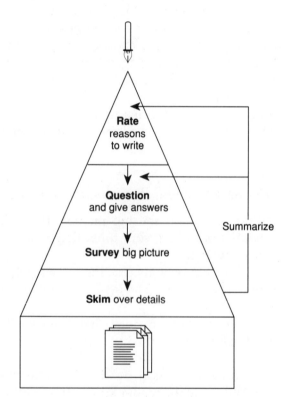

© 1991 by K. Alesandrini.

• Use labels and subheads as you write to keep the reader focused on how each section relates to the key questions or issues.
• State answers in concrete terms. Specific examples draw a picture in the reader's mind and avoid misunderstanding. Action verbs and familiar terminology enliven answers so readers can understand and remember the point.

3. Survey the Big Picture

Survey the big picture as you begin writing, then communicate it to the reader. The point of surveying is to communicate a sense of the whole. Give the general theme and the basic organization.

• Graphically depict the main points whenever possible. "A picture is worth a thousand words," goes the old saying. Survey a complex topic by illustrating it with an appropriate visual. This holds true not only for concrete topics but also for abstract ideas. Showing a schematic diagram or chart to explain a complex relationship can significantly reduce the amount of verbiage you'll need. And readers are more likely to understand and remember your message.

• Link ideas to the bigger picture with your chart or diagram. When you look at a road map, you can quickly see how points of interest relate to one another. Like a road map, linkages between main ideas will help your readers gain a bird's-eye view of the topic. The chart to illustrate this Action Plan shows how these five points tie together for top-down writing.

• Organize your thoughts with a visual outlining technique such as mind mapping or clustering. The problem with traditional, verbal outlining is that it limits your thinking. There must be a *"B"* if you have an *"A."* After trying mind mapping for the first time, seminar participants describe it as more efficient, natural, and free-flowing. Clustering is described in the popular book *Writing the Natural Way* by Gabriele Rico. "Clustering is a nonlinear brain-storming process," she explains, "akin to free association." Others have referred to visual outlining as "mind mapping," because it allows you to diagram ideas much the way they're organized in your mind. I used to create mind maps with paper and pencil, but now I use computer tools since they let me stay focused on my ideas, not the drawing process.

4. Skim Over the Details

Include details only as they are needed to support main points and answer the questions you have posed. Keep it simple and to the point. As Winston Churchill said, "Short words are best, and the old words when short are best of all."

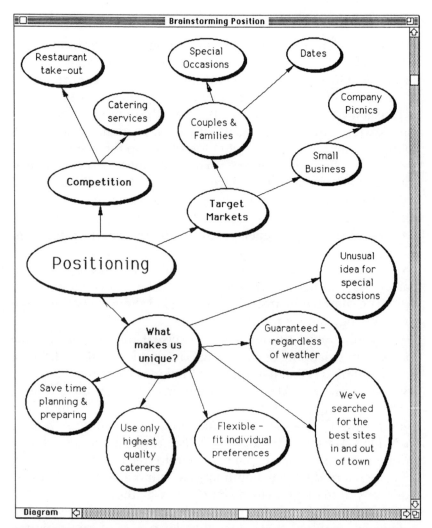

Links in this mind map remain connected when rearranged using *Inspiration* from Ceres Software.

• Enumerate details. One way to handle details is to list the particulars. Use either numbered or unnumbered lists headed by bullets or other markers.

• Dictation can save you time, because the average person writes only 20–30 words per minute longhand but talks up to 180 words per minute. With today's more conversational style of writing, dictation works well as long as you follow a written outline or map to keep an eye on the big picture as you talk.

• Reduce explanations to the bare minimum. Shorten lengthy material by converting clauses to phrases and phrases to one-word modifiers.

• Eliminate all nonessential words. Avoid using pidgin English, but try to eliminate all nonessential articles, connectors, phrases, and clauses. Use words as "reminders" rather than detailed explanations; assume the reader can rely on contextual cues for understanding.

• Provide supporting material in an attachment or appendix if needed rather than include detail within the body of the correspondence.

• Refer the reader to additional resources and tell where they can be found.

5. Summarize Periodically, Conclude with a Wrap-Up

Since readers don't always summarize for themselves as they read, it helps to summarize for them.

• Unify points into a memorable acronym or term. For example, the acronym CLEAR unifies five steps to better deskmanship.

• Summarize with an analogy. Relating new ideas to something familiar helps the reader retain your points.

Mastering the RQ3S writing method gives you the advantage of a bird's-eye view. You're left feeling on top of your workload rather than buried by it.

SELF-TEST: Is Business Communication Getting You Down?

Find out whether reading and writing are consuming more of your time and energy than they're worth.

	Yes	No
1. Do you rely primarily on a daily newspaper to stay up with important news events that could impact your business?	____	____
2. Do you read more than one daily newspaper?	____	____
3. Do you subscribe to an "industry watchdog" publication that provides an overview of your industry?	____	____
4. Do you refer to publications in your organization's library when needed rather than subscribe?	____	____
5. Do you read summaries and digests of information when they're available?	____	____
6. Do you use a clipping service, online search service, or other means to "delegate" your reading?	____	____
7. Do you survey and skim for main ideas rathr than read word for word?	____	____
8. Do you write an executive summary when you prepare lengthy reports?	____	____
9. Do you sit down and begin the writing process with the first paragraph?	____	____
10. Do you conclude memos and letters by making a request or telling the reader what you'd like him or her to do?	____	____

Scoring

Questions 1, 2, 9, 10: 1 point for each "yes."
Questions 3–7: 1 point for each "no."

What Your Score Means

Score of 0   Great! You're handling reading and writing requirements well.

Score of 1–5   You can improve your efficiency and effectiveness by skimming the chapter for

> new ideas and following the two Action Plans.

Score of 6–10   Business communication is getting you down! Begin with the Action Plan for better reading. Then apply your new reading skills to read and study this chapter.

Effective communication is essential for good management. Yet many managers continue to bury their messages in jargon and wordiness. In Britain, an organization called the Plain English Campaign is trying to shame individuals and organizations into streamlining their communications. Perpetrators of gobbledygook are goaded into doing better by receiving Golden Bull awards.

Every white-collar worker—whether a CEO, middle manager, or technical worker—is in the communications business. Learn to communicate well, and you increase your chance of success in the information age. Rely on the traditional approaches to business communication and you risk your very survival and that of your organization.

A Personal Postscript

I have not always kept an eye on the big picture. Back in 1978 and again in 1980, unusually heavy rains triggered land movement that damaged my home in the hills of Malibu. Although the media often stereotypes Malibu homeowners as wealthy actors and beach dwellers, the majority are not. At one time, Malibu property was inexpensive and one of the few areas adjacent to metropolitan Los Angeles where one could escape the bustle of city life while staying close enough to enjoy its many advantages.

On Thursday, March 9, 1978, a UPI photo of my neighbor putting his whole arm into the big crack in our street ran in my Midwestern hometown paper, *The Pekin Daily Times*. Later in the year, a few weeks before Christmas, a letter came from the county engineer declaring our little neighborhood of 20 or so homes a "Potential Landslide" area. And slide it did. Luckily, no one was injured and only two homes were a total loss, although a number were lost in adjacent neighborhoods.

For years afterwards, we had to endure a slow but relentless slide. Many got fed up with the broken pipes, shifting foundations, and ever-widening cracks in walls and ceilings. The topic of conversation centered around, "What's your new damage *this* week?" Commutes to work often were interrupted by road closures and repairs. A number of homeowners moved to "town" (into the city) to escape it all. I was one of those escapees.

I put my house up for rental, but it sat empty for nearly six months, which confirmed my view that nobody in their right mind would want any part of such a disaster. Then along came two people new to the area, who offered to rent the property. Before signing the contract, I wanted to meet them, so one Sunday afternoon I drove out to Malibu for our meeting.

The pair wanted to impress me that they would make good tenants. They told of their plans to plant flowers next to the patio off the living room. They explained how they would open the front door and living room slider for cross ventilation and bring patio furniture to take advantage of the ocean view. In short, they shared their vision—their big picture—of what having the property meant to them: a quiet, peaceful retreat close to nature to balance their stressful workdays.

As I listened to their plans, I slowly began to regain my own vision of what Malibu once meant to me. I had left the area some years earlier to work in the Midwest for five years. Missing the mild ocean air one winter's day, I wrote a song about Malibu. The words of that song filled my thoughts as I drove back to my city apartment, which was on a noisy street that reeked of car fumes: *The sea's callin' to my soul from the Malibu, saying 'Come back, come back.'*

Not surprisingly, I did move back. But this time, I brought a renewed vision with me. Today, the cracks are still with us, but we work around them and see "ground creep" as a small part of the big picture. Luckily, someone else's vision substituted as my "new eyes," and I retrieved this opportunity before it got away.

I think a big reason we fail to respond to opportunities or challenges in business is that we lack the needed vision—at least, until our competitor parades what we missed in front of our eyes, or a crisis waves a red flag under our noses. We don't see how sales could be improved until a competitor goes out and does it—and gains a competitive advantage in the process.

"The crisis you have to worry about most," according to former U.S. ambassador to Japan Mike Mansfield, "is the one you don't see coming." When the first Xerox machine came out in the 1950s, the profitable carbon paper companies did not see the enormous challenge it represented. Their sales and profits actually went up for a time as the public gained an awareness of the value of having copies of documents. But then the industry collapsed, and the carbon paper companies perished since they saw the big picture too late to secure their own survival.

"Of the top guys in the electronics industry back in 1955," points out Richard Foster of McKinsey & Co., "not one was a major player in electronics in 1980. Their skills, their whole way of thinking—'How can we make better vacuum tubes?'—led them to oblivion."

Companies who ignore the big picture suffer in the end. In contrast, those like McDonald's, who respond to changes not long after they appear on the horizon, survive and prosper. McDonald's executives realized the threat posed by changing consumer condemnation of companies that don't recycle or are not environmentally responsible. They responded by replacing the polystyrene burger boxes with paper wrappers.

Envisioning the possibilities is the first step to realizing them. Mike Vance, creative director at Walt Disney Studios, tells this story: "Soon after the completion of Disney World someone said, 'Isn't it too bad Walt Disney didn't live to see this.' I replied, 'He did see it—that's why it's here!'" Unfortunately, not everyone has the creative genius and vision of Walt Disney. Fortunately, everyone can benefit from the vision of other people's "new eyes."

Survival in the information age means getting outside of your own industry so that you can work more successfully within it. Survival means plugging into the collective wisdom, which elevates your own knowledge, which advances the collective wisdom, which further elevates your knowledge, which advances. . . . Well, you get the picture—the big picture—which should propel you and your business successfully into the 21st century.

Now, to complete the connection, I would like to hear from you. Please write me at my address or in care of Business One Irwin with your results from applying the ideas in this book. We welcome ideas and suggestions for any future editions. It's your chance to contribute something back to the collective wisdom.

Endnotes

Page 1: 750M printouts
"Imageprocessing," *VarBusiness*, August 1990, p. 28.

Page 1: 1.6 trillion pieces; 36 hr stacked up & 90 min to handle
USA Today, June 18, 1991, p. A1.

Page 1: office paper 2X GNP
"Paper, Once Written Off, Keeps a Place in the Office," *The New York Times*, July 7, 1990, p. 1.

Page 1: 60 billion faxes
"Fax Boards Are Supplement, Not Fax Machine Replacement," *Crain's New York Business*, April 29, 1991, p. 26.

Page 2: 12.5 E-mail users
Electronic Mail Association, letter to author, August 21, 1991.

Page 2: 1.2 billion messages AT&T Easylink Services, letter to author, December 4, 1991.

Page 2: 50% behind schedule
Dartmouth Research for AED Custom Training, reported in *Training & Development*, September 1991, p. 21.

Page 2: 50,000 books Grannis, C.B., "1990 Facts and Figures," *Publishers Weekly*, March 8, 1991, p. 36.

Page 2: 11K periodicals
Magazine Publishers of America, letter to author, September 26, 1991.

Page 2–3: Roger E. Herman, Herman Associates, telephone conversation with author, May 29, 1991.

Page 4: Daniel Burrus, president, Burrus Research, Inc., telephone conversation with author, August 23, 1991.

Page 4: "Information Visualizer"
Xerox PARC Information Center, letter to author, October 24, 1991; "An Easier Interface," *Byte Magazine*, February 1991.

Page 4–5: J. Kotter, "What Effective General Managers Really Do," *Harvard Business Review,* November–December 1982, pp. 156–67.

Page 5: Ryal Poppa, CEO, Storage Technology, telephone conversation with author, September 26, 1991.

Page 5: Beth Battram, Business One Irwin, telephone conversation with author, October 16, 1991.

Page 5: Jim Fitzgerald, manager, technical support, Borland, telephone conversation with author, September 20, 1991.

Page 6: "No company can duplicate . . ."
Jim Fitzgerald, manager, technical support, Borland, telephone conversation with author, September 20, 1991.

Page 6: "lack of collaboration"
Dartmouth Research for AED Custom Training, reported in *Training & Development,* September 1991, p. 20.

Page 6: Xerox study
T. A. Stewart, "Brainpower," *Fortune,* June 3, 1991, p. 50.

Page 6–7: C. H. House and R. L. Price, "The Return Map: Tracking Product Teams," *Harvard Business Review,* January–February 1991, pp. 92–100.

Page 8: Brenda Sumberg, director of quality, Motorola University, telephone conversation with author, October 9, 1991.

Page 8: Xerox researcher
M. Schrage, "Computer Tools for Thinking in Tandem," *Science Magazine,* August 2, 1991, p. 507.

Page 8: Small Group Collaborative Technology, EDS Center for Machine Intelligence, letter to author, October 14, 1991.

Page 8: 71 & 91 90 Savings
Post, B.Q., "Building the Case for Group Support Technology."

Page 8: J. Kotter, "What Effective General Managers Really Do," *Harvard Business Review,* November–December 1982, p. 164.

Page 8–9: Lewis Smith, president, Allied-Signal Aerospace, Kansas City Division, interview with author, Kansas City, Missouri, October 10, 1991.

Page 9: Daniel Burrus, president, Burrus Research, Inc., telephone conversation with author, August 23, 1991.

Page 9: publishing exec
Beth Battram, Business One Irwin, telephone conversation with author, October 16, 1991.

Page 10: T. Teal, "Service Comes First: An Interview with USAA's Robert F. McDermott," *Harvard Business Review,* September–October 1991, p. 120.

Page 11: 11-minute segments
USA Today, June 18, 1991, p. A1.

Page 11: 250 words per minute
R. Flesch, *The Art of Readable Writing* (New York: Macmillan, 1962), p. 181.

Page 11: 145 pages/week
D. Booher, *Cutting Paperwork in the Corporate Culture* (New York: Facts on File), p. 105.

Page 11: how time is spent
H. Poppel, "Who Needs the Office of the Future?" *Harvard Business Review,* November–December 1982, pp. 146–56.

Page 12: overload symptoms
Beyond the Paperless Office, Compugraphic Corp., White Paper, January 1985.

Page 12: R. Wurman, *Information Anxiety* (New York: Doubleday, 1989).

Page 12: J. Naisbitt, *Megatrends: Ten New Directions Transforming Our Lives* (New York: Warner Books, 1982), p. 17.

Page 13–14: D. Devoss, "If You Think You're Overloaded . . . 10 Southern Californians and Their Information Age Strategies," *Los Angeles Times Magazine,* January 22, 1989, pp. 14–17.

Page 15–16: Jim Manzi, CEO, Lotus Development Co., letter to author, August 31, 1990; R. Nelson, "Graphics: The Wretched Excess," *Personal Computing,* February 1990, p. 50.

Page 16: G. Blake, "Graphic Shorthand as an Aid to Managers," *Harvard Business Review,* March–April 1978, pp. 6–12.

Page 18: R. P. Gandossy and R. M. Kanter, "Kanter on Management," *Management Review,* March 1986, p. 14.

Page 18: "The Global Service 500," *Fortune,* August 26, 1991, p. 166.

Page 18: Productivity Figures
R. Henkoff, "Make Your Office More Productive," *Fortune,* February 25, 1991, p. 72.

Page 19: 12 percent productivity increase
H. W. Smith, "Productivity Pyramid," *How to Use the Franklin Day Planner,* Tape 2 Side B (Franklin International Institute, 1988).

Page 19: save 1½ hr per day
M. Babcock, director of marketing, Time/Design, telephone conversation with author, August 28, 1991.

Page 19: Daniel Burrus, president, Burrus Research, Inc., telephone conversation with author, August 23, 1991.

Page 19–20: M. Babcock, director of marketing, Time/Design, telephone conversation with author, August 28, 1991.

Page 22: Daniel Burrus, president, Burrus Research, Inc., telephone conversation with author, August 23, 1991.

Page 23: D. Steinberg, "Your PC Wish List," *PC Computing,* May 1991, pp. 155–59.

Page 25: Lotus Agenda Quick Start (Cambridge, Mass.: Lotus Development Corp., 1990), pp. 1–5.

Chapter 2 Notes

Page 32: survey of time pressure
"Time at a Premium for Many Americans," *The Gallup Poll Monthly,* November 1990, p. 43.

Page 32: waiting in line
D. Danbom, "Get in Line," *American Demographics,* May 1990, p. 11.

Page 32: wasted time
Michael Fortino, reported by Dan Sperling, "Study: Time's a Wasting," *USA Today,* June 6, 1988.

Page 32–33: McKinsey study
D. G. Reinertsen, "Product Races," *Electronic Business,* July 1983; McKinsey & Co, letter to author, November 1, 1991.

Page 33: 85 percent of executives
R. D. Pasquariello, *The Almanac of Quotable Quotes from 1990* (Englewood Cliffs, N.J.: Prentice Hall, 1991), p. 241.

Page 33: hours worked, $50K+
"Time at a Premium for Many Americans," *The Gallup Poll Monthly,* November 1990, p. 46.

Page 33: hours worked, entrepreneurs
"Time Management," *INC.,* November 1990, p. 144.

Page 33: J. Kotter, "What Effective General Managers Really Do," *Harvard Business Review,* November–December 1982, pp. 156–67.

Page 34–35: H. W. Smith and L. Vermillion, *6 Steps to Building the Productivity Pyramid* (Franklin International Institute, 1989).

Page 37: $350M Ford Edsel
D. Abadoher, *Iacocca* (New York: Macmillan, 1982), p. 137.

Page 37: $130 billion advertising
J. W. Wright (ed.), *The Universal Almanac 1991* (Kansas City: Andrews and McMeel, 1990), p. 235.

Page 37–38: R. M. Kanter, "Transcending Business Boundaries: 12,000 World Managers View Change," *Harvard Business Review*, May–June 1991, pp. 151–64.

Page 38: Wal-Mart
P. Sellers, "Does the CEO Really Matter?" *Fortune*, April 22, 1991, p. 82.

Page 39: Harley-Davidson
J. Stoltenberg, "See the Big Picture? Now Show Your Staff," *Working Woman*, April 1990, p. 85.

Page 39–40: A. Farnham, "The Trust Gap," *Fortune*, December 4, 1989, pp. 56–78; Forum Corp., *Corporate Strategy Study* (Boston, Mass.: The Forum Corp. of North America, 1989), p. 3.

Page 40: magazine editor
J. Stoltenberg, *Working Woman*, April 1990, p. 85.

Page 40: American Express
J. Stoltenberg, *Working Woman*, April 1990, p. 84.

Page 40: R. Kelley, "Poorly Served Employees Serve Customers Just As Poorly," *The Wall Street Journal on Managing* (New York: Doubleday, 1990), p. 71.

Page 41: Wally "Famous" Amos and G. Amos, *The Power In You* (New York: Donald Fine, 1988), p. 137.

Page 41: R. N. Bolles, *The 1991 What Color Is Your Parachute?* (Berkeley, Cal.: Ten Speed Press, 1991), p. 390.

Page 42: "American Management Operates," *Industry Week*, May 20, 1991, p. 26.

Page 42: Former CEO Canion
"How Top Managers Manage Their Time," *Fortune*, June 4, 1990, p. 250.

Page 42: Treleaven
J. Treleaven, V.P. and Partner, Battelle Venture Partners, telephone conversation with author, February 17, 1992.

Page 43: CEOs Burns, Sullivan, Rosenshine
"How CEO's Manage Their Time," *Fortune*, January 18, 1988, pp. 88–97.

Page 43: CEO Young
B. Dumaine, "How Managers Can Succeed through Speed," *Fortune*, February 13, 1989, p. 58.

Page 43: CEO Kurtzig
S. L. Kurtzig, *CEO: Building a $400 Million Company from the Ground Up* (New York: Norton, 1991), p. 57.

Page 44: S. Covey, *The 7 Habits of Highly Effective People* (New York: Simon & Schuster, 1989), pp. 98–99.

Page 45: E. C. Bliss, *Getting Things Done* (New York: Bantam, 1978), p. 40.

Page 47: Covey, p. 161.

Page 49–50: C. Panza, *Picture This . . . Your Function, Your Company* (Convent Station, N.J.: CMP Associates, 1991), p. 14.

Page 51: C. Panza and R. I. James, "Achieving Business and Team Building Objectives through a Single Program," *Performance & Instruction,* September 1991, p. 11.

Page 51: Carol Panza, management consultant, CMP Associates, telephone conversation with author, October 4, 1991.

Page 52: Brenda Sumberg, director of quality, Motorola University, telephone conversation with author, October 9, 1991.

Page 52: telephone company process map
C. Panza, *Performance & Instruction,* January 1989, pp. 10–17.

Page 57: survey of complaints
The Atlanta Journal and Constitution, September 4, 1990, p. D2.

Chapter 3 Notes

Page 58: 35–42 hours
"Buried by Paperwork?" *Los Angeles Times,* January 11, 1991, p. D5.

Page 58: 90 minutes
USA Today, June 18, 1991, p. A1.

Page 59: growth of paper 2x GNP
"Paper, Once Written Off, Keeps a Place in the Office," *The New York Times,* July 7, 1990, p. 1.

Page 59: statistics on paper
American Paper Institute, "Image Processing," *VarBusiness,* August 1990, p. 28.

Page 59: statistics on junk mail
Time, November 26, 1990, pp. 62–68.

Page 59: 1.6 trillion
USA Today, June 18, 1991, p. A1.

Page 59: "Battle Looms over Renewal of Federal Paperwork Reduction Act," *Washington Post,* September 11, 1989, p. WB12.

Page 59: H. Poppel, "Who Needs the Office of the Future," *Harvard Business Review,* November–December, 1982, pp. 146–56.

Page 59–60: time spent reading/writing
D. Booher, *Cutting Paperwork in the Corporate Culture* (New York: Facts on File, 1986), p. 109.

Page 60: "Health Care Paperwork Called Waste," *Washington Post*, May 2, 1991, p. A1.

Page 60: small business survey
The Atlanta Journal and Constitution, September 4, 1990, p. D2.

Page 60: "Congress's Uncontrollable 'In' Basket," *The Wall Street Journal*, May 15, 1991, p. A14.

Page 61: 99% have fax
IDC Market Interface, August 31, 1991, p. 5.

Page 61: "Fax Boards Are Supplement, Not Fax Machine Replacement," *Crain's New York Business*, April 29, 1991, p. 26.

Page 61: CEO Ounjian
E. Conlin, "The Daily Sales Report," *INC.*, January 1991, p. 73.

Page 62: Jim Fitzgerald, technical support manager, Borland Inc., telephone conversation with author, September 20, 1991.

Page 62: C. Blickenstorfer, "Where Does All the E-mail Go?" *Computerworld*, July 1991.

Page 62: 67% have E-mail
IDC Market Interface, August 31, 1991, p. 5.

Page 63: E-mail at DuPont
EMA Update 6, no. 1, 1991, p. 1.

Page 64: E-mail growth
Electronic Mail Association, letter to author, August 21, 1991; Eric Arnum, editor, *Electronic Mail & Micro Systems EMMS*, telephone conversation with author, August 21, 1991.

Page 64: J. Hirsch, "Flood of Information Swamps Managers," *The Wall Street Journal*, August 12, 1991, p. B1.

Page 65: E-mail at Hughes
"Mail Messages that Cut Red Tape," *Datamation*, October 15, 1990, pp. 47–50.

Page 66: Jim Fitzgerald, telephone conversation with author, September 20, 1991.

Page 67: "Reaping the Benefits of Financial EDI," *Management Accounting*, May 1991, pp. 39–44; Eric Arnum, editor, *Electronic Mail & Micro Systems EMMS*, telephone conversation with author, August 21, 1991.

Page 67–68: Parrish-Keith Simmons
"How to Survive and Thrive with EDI," *Industrial Distribution*, September 15, 1991, p. 37.

Page 68: Image processing growth
International Data Corporation, reported in *VarBusiness*, August 1990, p. 28.

Page 68: Small percentage exclusively on computer
"Image Processing: Its Role Beyond Check Processing," *EFT Report*, March 19, 1990, p. 1.

Page 68: Motorola catalog
"Paperless Office Revisited," *Electronic Buyers News*, September 3, 1990, p. 27.

Page 69: T. Teal, "Service Comes First: An Interview with USAA's Robert F. McDermott," *Harvard Business Review*, September–October 1991, p. 119.

Page 69: 98% reduction
J. Jakubovics, "The Care and Weeding of Office Records," *Management Solutions*, September 1988, p. 5.

Page 70: "Paperless Airplanes," *Tooling and Production*, January 1990, p. 34.

Page 71: 120 billion new sheets filed
J. Lawlor, "Orderliness . . . Is Power," *USA Today*, June 18, 1991, p. 1A.

Page 71: Filing statistics
Quill Business Library, *How to File and Find It* (Lincolnshire, Ill.: Quill Corp., 1989).

Page 72: Daniel Burrus, president, Burrus Research Inc., telephone conversation with author, August 23, 1991.

Page 76: Harvey Mackay, *The Harvey Mackay Rolodex Network Builder* (Rolodex Corporation, 1990), p. 34.

Page 78: Lou Filippo, "Coach," Sales and Management Development Co., telephone conversation with author, June 18, 1991.

Page 79: CEOs Crandall, Gill
"How Top Managers Manage Their Time," *Fortune*, June 4, 1990, pp. 88–97.

Page 79: Hogan B. Baird, marketing manager, Portable Computing Group, Lotus Development, letter to author, February 12, 1992.

Page 79: CEO Ounjian
E. Conlin, "The Daily Sales Report," p. 73.

Page 79: 45 minutes a day
J. Lawlor, "Orderliness . . . Is Power," p. 1A.

Page 80: clear desks
S. Winston, *The Organized Executive: New Ways to Manage Time, Paper, and People* (New York: Norton, 1983), p. 35.

Page 83–85: "What a Difference a Desk Makes," *Working Woman,* April 1990, p. 83. Reprinted with permission from *Working Woman* magazine. Copyright © 1990 by WWT Partnership.

Chapter 4 Notes

Page 91: fiber optics figures
G. Gilder, "Into the Telecosm," *Harvard Business Review,* March–April 1991, pp. 150–62.
Page 92: Visual channel
Data Management, July 1981, p. 30.
Page 93: G. Gilder, "Into the Telecosm," *Harvard Business Review,* March–April, 1991, pp. 150–62.
Page 93: public utility exec quote
"If You Think You're Overloaded," *Los Angeles Times Magazine,* January 22, 1989, p. 15.
Page 93: Information growth
J. Naisbitt, *Megatrends: Ten New Directions Transforming Our Lives* (New York: Warner Books, 1982), p. 16.
Page 93: 50,000 books
C. B. Grannis, "1990 Facts & Figures," *Publishers Weekly,* March 8, 1991, p. 36.
Page 93: Magazine Publishers of America, letter to author, September 26, 1991.
Page 93: member of Congress
J. Hirsch, "Flood of Information Swamps Managers," *The Wall Street Journal,* August 12, 1991, p. B1.
Page 93–94: J. C. Wetherbe, "Executive Information Requirements," *MIS Quarterly,* March 1991, pp. 54–55. Reprinted by permission of the *MIS Quarterly,* Volume 15, Number 1, March 1991. Copyright © 1991 by the Society for Information Management and the Management Information Systems Research Center at the University of Minnesota.
Page 94: #1 barrier to service
R. D. Pasquariello, *The Almanac of Quotable Quotes from 1990* (Englewood Cliffs, N.J.: Prentice Hall, 1991), p. 242.
Page 95: CEO M. Booth
"Brainpower," *Fortune,* June 3, 1991, p. 44.
Page 96: C. C. Gould and K. Pearce, *Information Needs in the Sciences: An Assessment* (Mountain View, Cal.: The Research Libraries Group, 1991), p. 76.

Page 96: zoo director quote
Los Angeles Times Magazine, January 22, 1989, p. 17.

Page 96: 3M Technical Forum
J. Diebold, *The Innovators: The Discoveries, Inventions, and Breakthroughs of Our Time* (New York: Dutton, 1990), p. 86.

Page 96: Kodak
Business Week, Quality 1991, p. 152.

Page 96–97: Amoco teams
R. Henkoff, "Make Your Office More Productive," *Fortune,* February 25, 1991, p. 84.

Page 97: speed at GE, AT&T, Navistar
B. Dumaine, "How Managers Can Succeed through Speed," *Fortune,* February 13, 1989, pp. 54–59.

Page 99–102: C. H. House and R. L. Price, "The Return Map: Tracking Product Teams," *Harvard Business Review,* January–February 1991, pp. 92–100.

Page 103–104: L. Butterfield, "Adopting Advanced Risk Evaluation Systems," *Construction Bulletin,* August 16, 1991, p. 10; Leslie Butterfield, vice president, Texim, telephone conversation with author, September 6, 1991.

Page 104: United Research survey
R. M. Isenhour, "Problems Implementing Time-Based Management," *Planning Review,* November–December 1990, p. 13.

Page 104: Midwest electronics firm
D. Valentino and B. Christ, "Time-Based Management Creates a Competitive Advantage," *Planning Review,* November–December 1990, p. 11.

Page 105: K. Alesandrini, "Pictures and Adult Learning," *Instructional Science* 13 (1984), pp. 63–77; "Imagery-Eliciting Strategies and Meaningful Learning, *Journal of Mental Imagery* 6 (1982), pp. 125–40.

Page 107: car dealership graphics
H. Takeuchi and A. H. Schmidt, "New Promise of Computer Graphics," *Harvard Business Review,* 1980, pp. 122–31.

Page 107: "Boom . . ."
W. Colby, "A Picture Is Worth a Thousand Printouts," *Infosystems* 31, no. 5 (1984), p. 82.

Page 108: B. Schwartz, "You Can Take Cigarette Ads Out of Television But You Can't Take Television Out of Cigarette Ads," *Media Industry Newsletter,* November 20, 1978, p. 7.

Page 108–109: Graphics aid software design
Robert Stanley, researcher, Cognos Inc., letters to author, March 12 and

March 14, 1990; T. Dudley, "Graphics in Software Design," *Computer Graphics World*, February 1986.

Page 109: Dr. Reatha King
J. Stoltenberg, "See the Big Picture? Now Show Your Staff," *Working Woman*, April 1990, p. 86.

Page 110: lawyers vs sculptors
J. E. Bogen, R. DeZure, W. D. TenHouten, and J. F. Marsh, "The Other Side of the Brain, IV: The A/P Ratio," *Bulletin of the Los Angeles Neurological Society* 37 (1972), pp. 49–61.

Page 110: D. Robey and W. Taggart, "Human Information Processing in Information and Decision Support Systems," *MIS Quarterly*, June 1982, pp. 61–73.

Page 111: "A Gestalt Completion Test." *Contributions to Education No. 48* (New York: Columbia University Teachers College, 1931).

Page 112: "Sharp Information Research Relief," *ASAP Magazine*, March–April 1989, p. 2.

Page 112: M. J. McCarthy, *Mastering the Information Age* (Los Angeles: Tarcher, 1991), p. 264.

Page 116: A. Glossbrenner, *How to Look It Up Online* (New York: St. Martin's Press, 1987).

Page 118: CompuServe Magazine, New Member, 1991, p. 23.

Page 119: CompuServe Magazine, new member, 1991, p. 25.

Page 122: J. S. McClenahen, "Nicolaus Copernicus, You're Needed Again," *Industry Week*, May 20, 1991, p. 58.

Page 123: Women in workforce
J. Naisbitt and P. Aburdene, *Megatrends 2000*, p. 245.

Page 124: Naisbitt and Aburdene, *Megatrends 2000*; S. Davis and B. Davidson, *2020 Vision: Transform Your Business Today to Succeed in Tomorrow's Economy* (New York: Simon & Schuster, 1991); C. R. Morris, *The Coming Global Boom: How to Benefit Now from Tomorrow's Dynamic World Economy* (New York: Bantam, 1990); J. H. Boyett and H. P. Conn, *Workplace 2000: The Revolution Reshaping American Business* (New York: Dutton, 1991); R. B. Tucker, *Managing the Future: 10 Driving Forces of Change for the '90s* (New York: Putnam, 1991).

Page 125: Kenneth Hey, managing partner, Inferential Focus, Inc., telephone conversation with author, September 30, 1991.

Page 127: time in meetings
H. Poppel, "Who Needs the Office of the Future," *Harvard Business Review*, November–December 1982, pp. 146–56.

Chapter 5 Notes

Page 128–29: *New Media Products,* February 1991, p. 4–6; 3M Visual Systems Division, letter to author, November 8, 1991.

Page 129: *% time in meetings*
H. Poppel, "Who Needs the Office of the Future," *Harvard Business Review,* November–December 1982, pp. 146–156; *AMA Management Review,* February 1982.

Page 129: *executive quote*
The 3M Meeting Management Team, *How to Run Better Business Meetings* (New York: McGraw-Hill, 1987), p. 3.

Page 129: *accountemps study*
Christian Science Monitor, July 10, 1990, p. 8.

Page 130: *McDonnell Douglas survey*
AV Video, January 1990, p. 76.

Page 130: *$37 billion*
"Making Meetings More Meaningful," *Washington Post,* July 1, 1990, p. H3 (1).

Page 130: *reasons meetings waste time*
Milo Frank, *How to Run a Successful Meeting in Half the Time* (New York: Simon & Schuster, 1989), p. 18.

Page 130: *Annenberg School of Communication study, USC*
New Media Products, pp. 4–6; 3M Visual Systems Division, letter to author, November 8, 1991.

Page 131: Minnesota Western, "Effective Meetings," *Visual Products Catalog,* Minnesota Western, p. 165; Reproduced by permission of and copyrighted by Minnesota Mining and Mfg. Co.

Page 132–134: *Wharton study*
The 3 M Meeting Management Team, *How to Run Better Business Meetings* (New York): McGraw-Hill, 1987; 3M Visual Systems Division, letter to author, November 8, 1991; Graphs on pages 133 and 135 reproduced by permission of and copyrighted by Minnesota Mining and Manufacturing Company.

Page 134–35: *New Media Products,* February 1991, pp. 4–6; 3M Visual Systems Division, letter to author, November 8, 1991.

Page 135: K. Alesandrini, *Power Graphics: Your Guide to Intelligent Presentations* (Lisle, Ill.: Pansophic Systems, 1989); *Master Graphics: Effective Overheads for Business Presentations* (Anaheim, Ca.: CALCOMP, 1988).

Page 136: *Eichenfield*
N. Semonski, Ventana Corp, letter to author, November 27, 1991.

Page 136: CEOs, Grove, Crandal
E. Calonius, "How Top Managers Manage Their Time," *Fortune*, June 4, 1990, pp. 250–62.

Page 136: Sandra L. Kurtzig and Tom Parker, *CEO: Building a $400 Million Company from the Ground Up* (New York: Norton, 1991), p. 141.

Page 136: Ryal Poppa, CEO, Storage Technology, telephone conversation with author, September 26, 1991.

Page 136: Fields S. Solomon, "Use Technology to Manage People," *INC.*, May 1990.

Page 137–38: Miles Babcock, director of marketing, Time/Design, telephone conversation with author, August 28, 1991.

Page 140: G. Walther, *Phone Power: How to Make the Telephone Your Most Profitable Business Tool* (New York: Berkley Books, 1986).

Page 142: Motorola Cellular Impact Survey (by Gallup Organization), (Arlington Heights, Ill.: Pan American Cellular Subscriber Group, May 1991). Telephone conversation with author November 12, 1991.

Page 145: E. Andrews, "Plugging the Gap Between E-Mail and Video Conferencing," *The New York Times*, June 23, 1991, p. F9.

Page 145: T. Reid Lewis, president, Group Technologies, Inc., telephone conversation with author, September 6, 1991.

Page 147: "Computer Tools for Thinking in Tandem," *Science Magazine*, August 2, 1991, p. 507.

Page 147: Small Group Collaborative Technology, EDS Center for Machine Intelligence, letter to author, October 14, 1991.

Page 149: F. Paul, "E-Mail Comes of Age," *Omni*, p. 35.

Page 149: Cheifet quote
CompuServe Magazine, New Member, 1991, p. 22.

Page 150: Alaska firefighters
"A Training Success Story," *Training & Development*, September 1991, p. 63.

Page 150: Harrison/Hofstra study
Washington Post, July 1, 1990, p. H3.

Page 153: Ryal Poppa, CEO, Storage Technology, telephone conversation with author, September 26, 1991.

Chapter 6 Notes

Page 158: Stata quote
T. Stewart, "Brainpower," *Fortune*, June 3, 1991, p. 60.

Page 158: C. Argyris, "Teaching Smart People How to Learn," *Harvard Business Review,* May–June 1991, pp. 99–109.

Page 159: The figures about the training industry all come from the American Society for Training and Development, letter to author, October 25, 1991.

Page 160: $12 billion for execs
"Back to School," *Business Week,* October 28, 1991, p. 102.

Page 296: 10,000 managers
Business Week, October 28, 1991, p. 109.

Page 161: Hovdey quote
Training & Development, August 1991, p. 29.

Page 161: "the company most committed . . ."
J. Naisbitt and P. Aburdene, *Megatrends 2000: Ten New Directions for the 1990's* (New York: Morrow, 1990), p. 242.

Page 161: Brenda Sumberg, director of quality, Motorola University, telephone conversation with author, October 9, 1991.

Page 161: T. Teal, "Service Comes First: An Interview with USAA's Robert F. McDermott," *Harvard Business Review,* September–October 1991, pp. 117–27.

Page 162–63: Ryal Poppa, CEO, Storage Technology, telephone conversation with author, September 26, 1991.

Page 163: Ingersoll-Rand
J. L. Sheedy, "Retooling Your Workers Along with Your Machines," *The Wall Street Journal on Managing,* ed., D. Asman (New York: Doubleday, 1990), pp. 76–79.

Page 163: quality survey
"Is Quality Priority One?" *Training,* September 1991, p. 88.

Page 163: Lewis Smith, president, Allied-Signal Aerospace, Kansas City Division, interview with author, Kansas City, Missouri, October 10, 1991. Operated for the U.S. Department of Energy by the Allied-Signal, Inc., Kansas City Division, under contract no. DE-ACO4-76DP00613.

Page 164: Anne Behrens, manager of organization development, Allied-Signal Aerospace, Kansas City Division, telephone conversation with author, October 18, 1991. Operated for the U.S. Department of Energy by the Allied-Signal, Inc., Kansas City Division, under contract no. DE-ACO4-76DP00613.

Page 164: Merck
J. H. Boyett and H. P. Conn, *Workplace 2000* (New York: Dutton, 1991), p. 323.

Page 164: top two training concerns
"Industry Report, 1991," *Training,* October 1991, p. 56.

Page 164: 80% labor
"Industry Report, 1991," *Training,* October 1991, p. 40.

Page 164: Diane Graham, director of training, Allied-Signal Aerospace, Kansas City Division, telephone conversation with author, October 18, 1991. Operated for the U.S. Department of Energy by the Allied-Signal, Inc., Kansas City Division, under contract no. DE-ACO4-76DP00613.

Page 166: "We have a process . . ."
Brenda Sumberg, telephone conversation with author, October 9, 1991.

Page 166: Penn State Univ
Business Week, October 28, 1991, p. 105.

Page 167: Ralph Adler, manager of computer reference, J.C. Penney Co., Inc., telephone conversation with author, October 16, 1991.

Page 168: "When managers look . . ."
Brenda Sumberg, telephone conversation with author, October 9, 1991.

Page 169: reading and communication percentages
A. Carnevale, L. Gainer, and A. Meltzer, *Workplace Basics: The Essential Skills Employers Want* (San Francisco, Cal.: Jossey-Bass Publishers, 1990).

Page 170: survey of learning how to learn
ASTD, *Workplace Basics: The Skills Employers Want* (U.S. Department of Labor, 1988).

Page 170: Bruce Petty
Visual Education Curriculum Project, *Pictures of Ideas: Learning through Visual Comparison and Analogy* (Canberra, Australia: The Curriculum Development Centre, Peacock Publications, 1981), p. 33.

Page 171–72: Carnevale, Gainer, and Meltzer, *Workplace Basics: The Essential Skills Employers Want* (San Francisco, Cal.: Jossey-Bass Publishers, 1990), p. 8.

Page 172: T. Stewart, "Brainpower," *Fortune,* June 3, 1991, p. 60.

Page 173: G. A. Miller, "The Magical Number Seven, Plus or Minus Two," *Psychological Review* 63 (1956), pp. 81–97.

Page 174: study of chess players
A. D. DeGroot, *Thought and Choice in Chess* (The Hague: Mouton, 1965).

Page 175: M. H. Erdelyi and J. Becker, "Hypermnesia for Pictures," *Cognitive Psychology* 6 (1974), pp. 158–71.

Page 175: K. Alesandrini, "Pictures and Adult Learning," *Instructional Science* 13 (1984), pp. 63–77; "Imagery-Eliciting Strategies and Meaningful Learning, *Journal of Mental Imagery* 6 (1982), pp. 125–40.

Page 177: R. Haavind, "Hypertext. The Smart Tool for Information Overload," *Technology Review,* November–December 1991, pp. 44–50.

Page 177–78: Federal Express 10% gain
"Training + Technology," *Training & Development,* September 1991, p. 50.

Page 178–80: Original test from: A. Richardson, "Verbalizer-Visualizer: A Cognitive Style Dimension," *Journal of Mental Imagery* 1 (1977), pp. 109–26. Modified based on findings of: J. R. Kirby, P. H. Moore, and N. J. Schofield, "Verbal and Visual Learning Styles," *Contemporary Educational Psychology* 13 (1988), pp. 169–84.

Page 181: top two training methods
"Industry Report, 1991," *Training,* October 1991, p. 50.

Page 181: Brenda Sumberg, Motorola University, telephone conversation with author, October 9, 1991.

Page 182: S. Yelon and W. Anderson, "Using Cartoons in Training Presentations," *Performance & Instruction,* November–December 1989, p. 21.

Page 184: Senge "The Learning Organization Made Plain," *Training & Development,* October 1991, p. 40.

Page 185: 1.5 to 2 hrs reading a day
A. P. Carnevale, *America and the New Economy* (Washington D.C.: American Society for Training and Development, 1991), p. 108.

Page 185: 92% value reading
"Signs Encouraging for an Upsurge in Reading in America," *The Gallup Poll Monthly,* February 1991, p. 43.

Chapter 7 Notes

Page 188: Roger E. Herman, Herman Associates, telephone conversation with author, May 29, 1991.

Page 190: C. C. Gould and K. Pearce, *Information Needs in the Sciences: An Assessment* (Mountain View, Cal.: The Research Libraries Group, 1991), p. 7.

Page 191: "The Power of Goals," *Success,* September 1991, p. 43.

Page 191: "Signs Encouraging for an Upsurge in Reading in America," *The Gallup Poll Monthly,* February 1991, p. 43.

Page 191: J. Calano and J. Salzman, *Careertracking* (New York: Simon & Schuster, 1988).

Page 191: Miles Babcock, director of marketing, Time/Design, telephone conversation with author, August 28, 1991.

Page 192: B. Greenberg, "Mass Media in the U.S. in the 1980s," in *The Media Revolution in America and in Western Europe,* [ed. E. Rogers and F. Balle], 1985.

Page 194: managers don't know what info they want
J. C. Wetherbe, "Executive Information Requirements: Getting It Right," *MIS Quarterly,* March 1991, pp. 50–66.

Page 195: Mellon bank book reports
J. Hirsch, "Flood of Information Swamps Managers," *The Wall Street Journal,* August 12, 1991, p. B1.

Page 195: D. Devoss, "If You Think You're Overloaded . . . 10 Southern Californians and Their Information Age Strategies," *Los Angeles Times Magazine,* January 22, 1989, p. 15.

Page 196: M. J. McCarthy, *Mastering the Information Age* (Los Angeles: Tarcher, 1991), pp. 172, 176.

Page 197: R. Haavind, "Hypertext. The Smart Tool for Information Overload," *Technology Review,* November–December 1991, pp. 46–47.

Page 202: M. C. Wittrock and K. Alesandrini, "Generation of Summaries and Analogies and Analytic and Holistic Abilities," *American Educational Research Journal* 27, no. 3 (Fall 1990), pp. 489–502.

Page 202: 8 percent
A. P. Carnevale, *America and the New Economy* (Washington, D.C.: American Society for Training and Development, 1991), p. 109.

Page 203: unnecessary reports
D. Booher, *Cutting Paperwork in the Corporate Culture* (New York: Facts on File), p. 59.

Page 203–205: J. S. Fielden and R. E. Dulek, "How to Use Bottom-Line Writing in Corporate Communications," *Business Horizons,* July–August 1984, pp. 24–30.

Page 206: computers and writing
E. Mendelson, *The New Republic,* February 22, 1988.

Page 210: G. L. Rico, *Writing the Natural Way: Using Right-Brain Techniques to Release Your Expressive Powers* (Los Angeles: Tarcher, 1983).

Page 213: Golden Bull awards
J. Bredin, "Say It Simply," *Industry Week,* July 15, 1991, pp. 19–20.

Personal Postscript

Page 215: G. C. Meyers, "Amending Murphy's Law," *SV Entertainment,* October 1991, pp. 17–18.

Page 216: Mike Vance quote
Working World, February 4, 1991, p. 6.

Resource Directory

To the best of my knowledge, all products and services listed in the Resource Directory and the book are trademarks, service marks, and/or registered trademarks of the companies producing the products or services.

Audio-Visual Resources

Learn more quickly and effectively with visualization and mediated instruction to see the big picture fast.

Alesandrini & Associates offers A-V resources to accompany this book, *SURVIVE Information Overload*. The *Video Overview* features live product demos and author interview. Other resources include *Resource Contacts*, a computer disk containing an extended version of this Directory with built-in contact manager for automatic dialing to phone and fax numbers, audiocassette programs, slide program and seminar kit, and train-the-trainer support. 23715 W. Malibu Rd., Suite 284, Malibu, CA 90265. 800/252-2665 or 310/456-7608.

CareerTrack offers audiocassette programs and video seminars as well as live seminars on a variety of business topics. Free full-color catalog available. CareerTrack, 3085 Center Green Drive, Boulder, CO 80301. 800/334-1018 or 303/440-7440.

Corporate Visions, from the producers of the cable television program *Window on Wall Street,* is a collection of video programs on a variety of topics for managers and entrepreneurs. Library-style packaging. Videos can be returned for credit within 30 days. Chesney Communications, 2302 Martin St., Suite 125, Irvine, CA 92715. 714/756-1905.

Newstrack Executive Tape Service includes two 90-minute tape summaries per month of major newspaper and magazine articles related to business. Tracks 150 publications. $249 per year. 700 Black Horse Pike, Suite 110, Blackwood, NJ 08012. 800/776-5771 or 609/232-6380.

Nightingale–Conant is the world's leading publisher of personal development audiocassette programs. Subjects include time management, memory enhancement, goal setting, decision making, stress reduction, and many more. The company has an extensive collection of personal development video programs. 7300 N. Lehigh Ave., Chicago, IL 60648. 800/525-9000 or 708/647-0300.

Calendar and Scheduler Software (See also Personal Information Manager and Project Management)

Keep track of personal and group schedules by computer so that you can see your activities in context.

CA-UpToDate is a personal/group information manager designed to help you manage time, personnel, and projects with diaries, resource parameters, and up to 65,000 schedules. Includes password protection and full network compatibility. Available through retailers for Windows. Computer Associates Int'l, Inc., 1 Computer Associates Plaza, Islandia, NY 11788. 800/225-5224 or 516/227-3300.

OnTime features a lifetime calendar, tickler alarm system, automatic rollover of uncompleted to-do list tasks, and key-word searching. DOS, $69.95; Windows, $129.95. Network versions available. May be purchased through software retailers. Campbell Services, Inc., 21700 Northwestern Highway, Suite 1070, Southfield, MI 48075. 800/345-6747 or 313/559-5955.

CD–ROM

Save money when you access large databases and other types of information on disc rather than pay high online fees that are charged by the minute. CDs allow you to search broadly to find the specific information you need.

COMLINE on SilverPlatter is a source for worldwide news on Japanese high-tech companies. It includes over 75,000 English-language summaries, abstracted from over 130 Japanese publications. Includes one disc, updated bimonthly. Single user, $995; Multi-user, $1,592. Available through SilverPlatter Information, Inc., 100 River Ridge Dr., Norwood, MA 02062. 800/343/0064 or 617/769-2599.

European Corporations CD includes the full and unedited text of the annual reports and financial statements for 3,000 European companies.

Most records contain commentary on performance, review of business climate, and more. Includes two discs, updated monthly. Single user, $6,750. Available through SilverPlatter Information, Inc., Norwood, MA 02062. 800/343-0064 or 617/769-2599.

General BusinessFile is a CD-ROM business reference product combining periodicals, newswires, company directories, and investment analyst reports. *Investext* is another optical disc reference product that includes company and research reports from 60 leading Wall Street firms. Information Access Company, 352 Lakeside Drive, Foster City, CA 94404. 800/227-8431 or 415/378-5200.

Lotus CD/Corporate is part of the *Lotus One Source* family of CD-ROM-based information products. Provides business news, statistics, and financial information through weekly, monthly, or quarterly CD-ROM and online updates. Lotus *Agenda* can retrieve and sort information from the CDs. Lotus Development Corp., 55 Cambridge Parkway, Cambridge, MA 02142. 800/544-5501 or 617/693-7833.

Predicasts F&S Index plus Text brings together over 1,000 trade, business, and government publications. Over 80 percent of the records feature full text or abstracts about companies, products, markets, and applied technology. Two discs, updated monthly. U.S. edition, $2,500; Int'l, $3,500. Available through SilverPlatter Information, Inc., 100 River Ridge Dr., Norwood, MA 02062. 800/343-0064 or 617/769-2599.

Clipping Services

Let several hundred pairs of eyes keep track of the big picture for you and receive timely information that impacts your business.

Burrelle's monitors 17,000 publications, major newswires, TV, and radio. The company offers: *NewsExpress* for news faxed to you by 9:00 A.M. every business day, *Broadcast Database* of TV programs, *Advertising Analysis*, and more. See also "Information and Research Services" section. Burrelle's Information Services, 75 E. Northfield Rd., Livingston, NJ 07039. 800/631-1160 or 201/992-6600.

Luce Press Clippings monitors 17,000 publications, major wire services, and television news. *AM Newsbreak* service faxes morning news. *Teleclips* service provides transcripts or tapings of TV stories; *Television Notification* notifies client if keyword is mentioned on TV. Reading charge for basic service is $210 monthly, plus $1.12 per clip. 42 South Center, Mesa, AZ 85210. 800/528-8226 or 602/834-4884.

Radio TV Reports tapes any type of material off-the-air that was broadcast on radio or television within the last 60 days and sends it to you on VHS or 3/4-inch tape for a fee, generally under $100 per program. Radio TV Reports, 41 East 42nd St., New York, NY 10017. 212/309-1400 or 213/466-6124.

Vu/Clip is the clipping-service companion of Vu/Text, the largest on-line newspaper database in the United States. Customer service reps take the keywords you want to track over the phone. Hits are mailed weekly or monthly for $10 per search, plus 75 cents per page. Vu/Text, 325 Chestnut St., Suite 1300, Philadelphia, PA 19106. 800/258-8080 or 215/574-4421.

Contact Management and Sales Automation (See also Personal Information Manager-PIM and Project Management)

Use this class of tailored database software to synthesize details related to people you contact by phone or mail.

ACT! is an award-winning contact management program that integrates a contact database, auto-dialer, scheduler, report generator, and full-featured word processor. Features faxing, electronic calendars, and more. $395 (PC version). Also Macintosh, palmtop PC, Penn-Windows and novice versions. Contact Software Int'l., 1625 Crosby Rd., #132, Carrollton, TX 75006. 800/365-0606 or 214/418-1866.

Address Book Plus maintains address information for printout in a variety of formats including popular organizer pages. Features built-in dialer, mail merge, bar code printing, automatic entry of common data, and more. Macintosh and DOS versions. Power Up Software, PO Box 7600, San Mateo, CA 94403. 800/851-2917 or 415/345-5900.

Business Contacts & Information Manager is a sales and client management tool featuring pop-up notepad, word processor, mail management, and auto dialing features. It allows users to file and retrieve client profiles, correspondence, vital data, histories of meetings, commitments, orders, and more. DOS version, $69.95; Windows version, $99.95. Disk-Count Software, 1751 W. County Rd. B, Suite 107, St. Paul, MN 55113. 612/633-2300.

Market Force Plus is an integrated sales and marketing automation system which maintains client database, handles field sales, telesales, direct marketing, and custom management reporting. Optional modules for corporate-wide sales automation. User customizable. Starts at $695 single, $2,495 LAN. Software of the Future, Inc., Box 531650, Grand Prairie, TX 75053. 800/766-7355 or 214/264-2626.

NameBank is a software companion to the Time/Design Management System allowing users to cross-reference and categorize names, addresses, and notes. Easily printed out, the reports fit right into the Management System. DOS, $95. Time/Design, 265 Main St., Agawan, MA 01001. 800/637-9942 or 413/789-9633.

TeleMagic autodials contacts, times each call, files notes for later, and prompts you with scripts for selling and answering questions. It generates mailing lists, labels, envelopes, and form letters. Unlimited support for first 90 days. $495, DOS; from $995 for network versions. PC, Mac, and others. Remote Control Int'l, 5928 Pascal Court, Suite 150, Carlsbad, CA 92008. 800/992-9952 or 619/431-4000.

Directories, Summaries, and Profiles

Add these resources to your big-picture reference shelf.

Communication Briefings is an eight-page monthly newsletter read by over 200,000 business and communication professionals worldwide. It covers advertising, PR, training, marketing, and more in concise language. $69 per year, U.S.; $79, Canada; $99, all other countries. Communication Briefings, 700 Black Horse Pike, Suite 110, Blackwood, NJ 08012. 800/888-2084 or 609/232-8286. Fax 609/232-8245

Executive Book Summaries service sends subscribers two or three eight-page summaries of top business books each month. Each summary is a distillation that preserves the content and spirit of the entire book. Annual subscription includes 35 summaries, $96. Soundview Executive Book Summaries, 5 Main Street, Bristol, VT 05443. 800/521-1227 or 802/453-4062. Fax 802/453-5062.

Hoover's Handbook annually profiles key companies in one-page overviews. *Hoover's Handbook of American Business* covers U.S. public and key private companies as well as nonprofits; *World Business* profiles key countries, regions, and top non-U.S. companies. $24.95 and $21.95. The Reference Press, 6448 Hwy. 290 E, Suite E-104, Austin, TX 78723. 800/486-8666 or 512/454-7778.

Lesko's Info-Power is a 1,086-page resource of over 30,000 free and low-cost sources of information for investors, businesses, etc. Audio and video programs available. $33.95 plus postage/handling. Information USA, PO Box E, Kensington, MD 20895. 800/32-LESKO or 301/942-6303.

The National Directory of Addresses and Telephone Numbers is phonebook size yet has over 230,000 listings for businesses, agencies, institutions, and more. Features 90,000 fax numbers, "City Profiles" for business travelers, yellow pages, and headquarters flagging. Also CD-ROM and custom editions. $59.95. Omnigraphics, Inc. Penobscot Building, Detroit, MI 48226. 800/234-1340 or 313/961-1340.

Electronic Data Interchange (EDI)

Eliminate paperwork altogether by exchanging computer data (formatted according to X12 or other standards) directly with your suppliers and other trading partners.

AT&T EDI solutions include network, systems, and support. Features message screening, multiple formats, mailboxes, multivendor connectivity, and more. *FreeForm Conversion* service translates EDI information to a human-readable format (fax, electronic mail, printout) for partners who don't have EDI systems. Contact your local AT&T Easylink Services representative or 800/325-8309.

expEDIte is IBM's family of EDI solutions that provide a full range of EDI offerings: mail and network services, data translation and management software, and consulting services for PCs, midrange and mainframe systems. IBM can help design, develop, and implement EDI systems and processes to gain a competitive edge. Contact your local IBM representative or the EDI Solutions Center at 800/284-5849.

Interconn is a complete EDI system for PCs running DOS. System price range, $2,000–$3,500. *Datatran* is a complete EDI system for PCs, minis, and mainframes running UNIX derivative operating systems. System price range $3,500–$23,000. Both systems support all major EDI standards and all major EDI mailbox services. St. Paul Software, 754 Transfer Rd., St. Paul, MN 55441. 612/641-0963.

STX12 for micro and mainframe computers supports a variety of data formats. Data may be entered via user-customized data entry screens or printed with customized reports. Also provides an application interface and unattended operation. Transfer can be direct or through any of the public networks. $2,295. Supply Tech, Inc., 1000 Campus Dr., Suite 200, Ann Arbor, MI 48104. 313/998-4023.

Electronic Mail (E-Mail)

Replace paper communications with electronic messaging.

AT&T Mail is one of the world's largest international public electronic messaging services and is also available on private networks via AT&T software. Messages may consist of text, graphics, spreadsheets, and computer files and may be delivered by U.S. mail, courier, telex, fax, satellite pagers, and voice systems. Contact your AT&T rep or 800/367-7225 in the U.S.; +908-658-6175, Africa or the Americas; 613-788-5815, Canada; +322-676-3737, Europe; 81-3-5561-3411, Japan; or +852-846-2800 in the Pacific Rim.

MCI Mail is an electronic mail service that users can access from any kind of computer by using basic communications software. MCI charges a $35 annual fee and bills by number of characters in the message: 1–500, 45¢; 501 to 2,500, 75¢; 2,501 to 7,500, $1; and every additional 7,500 characters, $1. MCI Mail, 1111 19th Street N.W., Suite 500, Washington, DC 20036. 800/444-6245 or 202/833-8484.

QuickMail is an E-mail software program which runs on your network. Comes with server and client software for Mac, DOS, and other machines included in the same package. Available in 1, 5, 10, 50, and 100 user packs. $99-$4,699. CE Software, 1801 Industrial Circle, West Des Moines, IA 50265. 800/523-7638 or 515/224-1995.

Electronic Forms—See Forms Software

Electronic Meeting Systems and Services (See also Groupware and Meeting Management)

Collaboratively *do* work in meetings rather than just *talk* about work by interacting through a networked computer system that facilitates brainstorming and group decision making.

GroupSystems provides a set of software tools to support a broad array of group activities such as electronic brainstorming, idea organization, voting, ranking, and more. Runs on DOS-based local area network (LAN), $42,500 for *Basic Toolset*. Meeting room rental or on-site portable set-up available. Ventana Corp., 1430 E. Ft. Lowell, Suite 301, Tucson, AZ 85719. 800/368-6338 or 602/325-8228.

TeamFocus reduces meeting time up to 40 percent. The software runs on a local area network (LAN) and features electronic brainstorming, vote tallying, ranking, issue analysis, idea organization, and more. Site license available. Also, 50 Team-Rooms may be rented across the United States and Canada through IBM branch offices. Contact your local IBM representative or the IBM TeamFocus Project Office at 301/640-5008.

VisionQuest supports face-to-face meetings in computer-equipped rooms with or without a facilitator. Operating on a local area network (LAN), it also allows dialogues between people at different times, and in different places. Features brainstorming aids, ranking, scoring, voting, and more. DOS. Collaborative Technologies Corp., 8920 Business Park Drive, Suite 100, Austin, TX 78759. 512/794-8858.

Environment (See Recycled Products and Environmental Issues)

Facsimile

Reduce paperwork and increase productivity through effective use of fax technology.

ABCfax allows callers to use touch-tones in response to voice prompts to request and receive faxes automatically, 24 hours a day. Works with *ABCvoice*, a PC-based voice mail, auto-attendant, and information retrieval system. Handles multiple lines simultaneously with up to 6 lines in one computer. Network compatible. Versicom Corp., 316 Regency Ridge, Dayton, OH 45459. 513/438-3700

AT&T Enhanced FAX enables user to send a fax to the AT&T network (via an 800 number) from a stand-alone fax machine. The fax can then be broadcast simultaneously to a thousand fax machines worldwide, held in the network for pickup (useful for travellers), or delivered at off-peak hours. Broadcast lists can be stored. Contact the local AT&T EasyLink Services rep or call 800/242-6005 (accessed via USA Direct from outside the United States).

FaxPump is a total fax/voice-mail system that allows callers to pick documents via touch-tone phone, then receive automatic fax response from unattended PC. Suitable for departments or small offices to handle inquiries. Includes voice-mail board, fax board, software, and accessories.

DOS, $1,295. 181 N. Central Ave., Campbell, CA 95008. 408/244-5600. Call 408/244-5773 for demo.

Hotelecopy Faxmail is a large public access fax network with over 2,000 locations nationwide including hotels, airports, and even convenience stores. Offers broadcast fax capability as well as other business services including phone dictation, language translation, telex, word processing, and more. Hotelecopy Faxmail Inc., 17850 NE Fifth Ave., Miami, FL 33162. 800/322-4448 or 305/651-5176.

TelePort enables Macintosh users to send and receive FAX documents. The fax unit connects directly to the Apple DeskTop Bus and allows users to switch from "Print" to "Fax" in the File menu, making faxing as easy as printing. It features autoanswer and autodial. Global Village Communication, Inc., 1204 O'Brien Drive, Menlo Park, CA 94025. 415/329-0700. Fax 415/329-0767.

Forms Software

Eliminate paperwork by filling out forms on computer.

F3 Forms Automation System includes *F3 Pro Designer*, a sophisticated forms design tool; *Design & Mapping*, a programming language; and *F3 Fill*, a freestanding electronic form filler. Professional design services and training are also available. *F3 Fill* runs on DOS, LAN, and Windows. BLOC Development Corp., 600 Fairway Drive, Suite 201, Deerfield Beach, FL 33441. 800/477-2562.

Informed for the Macintosh consists of *Designer* ($295) to create paper or electronic forms and *Manager* ($195 or $395 for four) to electronically fill them out and perform database functions. $395. Shana Corp., 105, 9650-20 Avenue, Edmonton, Alberta, Canada T6N 1G1. 403/463-3330. Fax 403/428-5376. AppleLink CDA0004.

JetForm Design is a Windows-based package for designing forms including those with bar codes. *JetForm Filler* lets you fill out the forms you've designed in either a Windows or DOS environment, and *Merge* takes information from existing databases as forms input. $495. JetForm Corp., 163 Pioneer Dr., Leominster, MA 01453. 800/267-9976 or 613/594-3026.

Free (or Nearly Free) Products & Services

How to File is a full-color 35-page booklet with photos, examples, and tips on filing supplies and systems. $4.95. Esselte Pendaflex Corp., Clinton Rd., Garden City, NY 11530. 516/741-3200.

How to File and Find It gives you 64 pages of solid how-to's including a section on micrographics. $3.95 a copy; $1.99 each for 50 or more. Quill Corp., 100 Schelter Re., Lincolnshire, IL 60069. 708/634-4800.

National Zip+4 Diskette Processing Service is a one-time free service offered by the U.S. Postal Service to update mailing lists. The service standardizes addresses and converts five-digit zip codes into nine-digit. Supply your MS-DOS list in Fixed ASCII format. National Address Information Center, Diskette Processing Services, 6060 Primacy Pkwy., Suite 101, Memphis, TN 38188. 800/238-3150.

Groupware (See also Electronic Meeting Systems and Services and Meeting Management)

Communicate and collaborate with others through groupware, the next step beyond E-Mail.

Aspects software for the Macintosh makes it possible for up to 16 widely dispersed people to work simultaneously on the same document and see each other's changes to the word processing, graphics, or other document in less than a second. Requires networked or modem-linked Macs. $299. Group Technologies, Inc., 1408 N. Fillmore Street, Suite 10, Arlington, VA 22201. 703/528-1555.

Instant Update software for the Macintosh lets groups collaborate on any word-processing document at different times and in different places. The program consolidates input from different users on the network automatically. One click shows an item's author and the time it was entered. Two users, $495; Five users, $995. ON Technology, Inc., 155 Second St., Cambridge, MA 02141. 617/876-0900.

Lotus Notes runs on local- and wide-area networks to allow integrated computer conferencing, broadcast dissemination, executive information systems, and mail-enabled applications. Includes a full-featured electronic mail subsystem. $62,500 for a minimum configuration and maintenance. Lotus Development Corporation, 55 Cambridge Parkway, Cambridge, MA 02140. 617/577-8500.

Information and Research Services

Let others find the facts or view the big picture for you.

Burrelle's Information Search Service retrieves information for clients who need fast answers. Burrelle's can access more than 300 databases and provide answers by phone, mail, or courier. Burrelle's Information

Services, 75 E. Northfield Rd., Livingston, NJ 07039. 800/631-1160 or 201/992-6600.

D&B Express allows non-subscribers to obtain a Business Information Report by phone for any company in the Dun & Bradstreet information base (9 million in U.S. and 7 million worldwide). Reports by phone, then fax or mail. $60 each on limited basis. Available to subscribers as *DUNS Dial*. D&B Information Services, N. America, One Diamond Hill Rd., Murray Hill, NJ 07974. 800/879-1362.

The FISCAL Directory of Fee-Based Information Services in Libraries. A 590-page notebook of libraries nationwide that offer information search and retrieval services for a fee. $29. FYI/County of Los Angeles Public Library, 12350 Imperial Highway, Norwalk, CA 90650. 800/582-1093.

Inferential Focus Briefing provides a view of the future in all areas of business and society worldwide. Partners of the firm cover 350 publications and infer important changes and opportunities, conveyed to clients in part through effective use of cartoons. Inferential Focus, 200 Madison Ave., New York, NY 10016. 212/683-2060.

Michigan Information Transfer Source will search and send you a photocopy of any available article for a $10 fee. University of Michigan, 106 Hatcher Graduate Library, Ann Arbor, MI 48109. 313/763-5060.

Mail Order

Many of the products and services in this directory are available directly from the manufacturer or service provider through mail order. Save time and hassles when you order direct, whether traditionally or electronically.

The Electronic Mall™ on the CompuServe Information Service offers thousands of products online. As a consumer, you can order everything from computer and office supplies to books and specialty items. As an advertiser, you can reach nearly a million prospects. CompuServe's Electronic Mall Group, 5000 Arlington Centre Blvd., Columbus, OH 43220. 800/848-8199 or 614/457-8600.

Staples, known as "the Office Superstore," discounts major office products including furniture, computer supplies, paper, filing supplies, fax machines, copiers, phones, and much more. Prices are deeply discounted (an average of 50 percent) from "catalog list" from top companies such as Canon, Hammermill, Hewlett Packard, Panasonic, Rolodex, Sony, 3M, and many others. Staples, Inc., PO Box 160, Newton, MA 02195. 800/333-3330 or 617/965-7030.

Meeting Graphics (See Presentation Graphics)

Meeting Management (See also Electronic Meeting Systems and Groupware)

Get more out of meetings yet put less time in by managing meetings more effectively.

Meeting Maker software for the Macintosh is a meeting scheduler and manager that lets you plan and schedule meetings and manage details like time, date, attendees, rooms, resources, and agendas. Includes personal calendar that can be printed in various appointment book formats. Five users, $495; 10 users, $895. ON Technology, Inc., 155 Second St., Cambridge, MA 02141. 617/876-0900.

Micrographics and Electronic Filing

Eliminate clutter and retrieve information more effectively by storing or referencing information electronically.

Canofile 250 is a desktop electronic filing system with built-in high-speed scanner than an automatically input 40 pages per minute. Odd-sized documents can be fed manually. The easy-to-use indexing system allows retrieval by name, number, date, visual symbol such as a logo, and more. $12,200. Canon U.S.A., Inc., One Canon Plaza, Lake Success, NY 11042. 516/488-6700 or 714/753-4000.

Canon CAR System is a computer-assisted retrieval system that lets users index and store documents on microfilm. Available for single user or up to 32 multiuser workstations in four departments. Single user, $15,000; Multiuser, $37,000, including computer, software, and reader printer. Canon U.S.A., Inc., One Canon Plaza, Lake Success, NY 11042. 516/488-6700 or 714/753-4000.

Online Databases (See also Online Services and Aides)

Tap into a world of information yet find just the detail you want quickly.

Business International combines 15 publications for managers of international and global operations. It tracks critical trends and analyzes what worldwide developments mean for international companies. It covers topics such as organization, strategic planning, manufacturing, R&D, trade, and much more. Available through Dialog (800/334-2564) from Business Int'l Corp., 215 Park Ave South, New York, NY 10003.

Company Intelligence, a combined directory and company news file, contains current addresses, financial, and marketing information as well as up to 10 recent news items of over 116,000 U.S. private and public companies. *Trade & Industry Index* provides indexing and, where available, full text from over 1,300 publications. Both available on Dialog (800/334-2564) from Information Access Company.

D&B Business Information Report provides perspective on a firm's operations and profitability based on the Dun & Bradstreet information base (9 million U.S. businesses and 7 million worldwide). Over one million financial statements on public and private firms. Available on subscription basis. D&B Information Services, N. America, One Diamond Hill Rd., Murray Hill, NJ 07974. 800/234-3867.

Japan Technology contains abstracts in English of articles from leading Japanese business and technical journals. It covers Japan's R&D, business strategies, and market trends. Most source articles are in Japanese. Translation services are available upon request. Database available through Dialog (800/334-2546) from SCAN C2C, Inc., 500 E. Street SW, Washington, DC 20024. 202/863-3850.

Magazine Index covers general-interest magazines on current affairs, politics, and cultural trends. In addition to its extensive indexing, *Magazine Index* contains the full text from over 530 popular U.S. and Canadian magazines. Available on Dialog (800/334-2564) from Information Access Company, 362 Lakeside Dr., Foster City, CA 94404. 800/227-8431 or 415/378-5249.

Online Services and Aides (See also Online Databases)

Search for important information you need broadly, then retrieve what you need with single-minded focus.

BRS Information Technologies provides menu-driven access to over 150 databases including *ABI/INFORM, PTS PROMT,* and the *Wilson Business Periodicals Index. BRS Colleague* option adds E-mail and newswire services. Databases cover a wide range of subjects. BRS Information Technologies, 8000 Westpark Dr., McLean, VA 22102. 800/955-0906 or 703/442-0900.

CompuServe® is the largest worldwide online information service to modem-equipped PC users with 1,500 databases and 300 special-interest forums including 225 high-tech companies that provide online computing support. One-time start-up fee of $50 plus charges for online time. CompuServe Inc., 5000 Arlington Centre Blvd., PO Box 20212, Columbus, OH 43220. 800/848-8199 or 614/457-8600.

Dialog provides access to over 425 online databases that can be searched by terms, dates, SIC codes, and more. *First Release* delivers up-to-the-minute news from the newswires to your PC. *Dialog Alert* automatically sends information on any topic once it becomes available in printed or electronic form. Dialog Information Services, Inc., 3460 Hillview Ave., Palo Alto, CA 94304. 800/334-2564 or 415/858-2700.

Pro-Search software helps you figure out how best to search for information *before* going online to keep costs down. Construct simple or complex searches off-line. $295 for the Macintosh, $495 for the IBM PC and compatibles. Optional *Pro-Cite* and *Biblio-Links* enable you to find and format bibliographies automatically. Personal Bibliographic Software, Inc., PO Box 4250, Ann Arbor, MI 48106. 313/996-1580.

TAPCIS saves time and money by keeping connect time to a minimum. *TAPCIS* allows users to download E-mail and forum messages for off-line reading. Support is available both online and through an 800 number. IBM PCs and compatibles, $79; OS/2 version, $99. Company offers 90-day money-back guarantee. Support Group Inc., PO Box 130, McHenry, MD 21541. 800/872-4768 or 301/387-4500.

Outlining and Diagraming Software

Keep track of your big-picture categories and convert abstract verbiage into clear diagrams.

allCLEAR turns written instructions into flowcharts automatically. You type your procedures, and the program's editor automatically does all the layout and drawing. Use any word processor, the software's editor, or existing spreadsheet, text, or database files. $299.95, DOS version. Clear Software, Inc., 385 Elliot Street, Newton, MA 02164. 800/338-1759 or 617/965-6755.

INSPIRATION is a visual "thought processor" for the Macintosh that lets you see your thoughts take shape in both visual and text forms. It helps you create outlines, mindmaps, diagrams, flowcharts, etc. $295.

Ceres Software, 2920 Southwest Dolph Court, Suite 3, Portland, OR 97219. 800/877-4292 or 503/245-9011.

TopDown is a top-rated charting program for the Macintosh designed to create flowcharts, procedures, graphic outlines, structure diagrams, and more. Lines automatically reroute when symbols are moved or resized. Charts can be created automatically from imported, typed outline. $345. Kaetron Software Corp., 12777 Jones Road, Suite 445, Houston, TX 77070. 713/890-3434. Fax 713/890-6767.

Personal Information Manager (PIM) (See also Contact Management and Project Management)

Turn your computer into one of the most important tools for surviving information overload by managing people, projects, and information.

Agenda performs a variety of functions: calendar scheduling, contact management, project management, and information sifting (sorting information from external sources such as CD-ROM and online services). It is fully customizable and allows users to examine information from different perspectives. $395. Lotus Development Corp., 55 Cambridge Parkway, Cambridge, MA 02142. 800/343-5414 or 617/577-8500.

IBM Current provides contact management, built-in calendar, and desktop organizer. Allows you to organize and retrieve text and graphical data. Easy to customize. Windows version, $395. IBM Desktop Software, 472 Wheelers Farms Rd., Milford, CT 06460. 800-IBM-2468.

TimeBase is a software companion to the Time/Design Management System designed to manage people, plans, and projects. By entering information once, the software automatically cross-references up to 14 screens and reports including charts. DOS, $195. Network version available. Time/Design, 265 Main St., Agawan, MA 01001. 800/637-9942 or 413/789-9633.

Presentation Graphics Products and Services

Enhance meeting effectiveness by using presentation graphics to clarify the message.

Autographix Overnight Slide Service creates 35mm slides and other media such as overheads, posters, and hard copy from users' PC and Mac computer files. Call for nearest imaging center. Autographix also offers equipment for producing slides, printed matter, and other graphics in-house. Autographix, Inc., 63 Third Ave., Burlington, MA 01803. 800/548-8558 or 617/272-9000.

Freelance Graphics for Windows is a full-featured presentation graphics package that features SmartMasters fill-in-the-blanks pages. The program also includes complete text and data charting, outlining capabilities, and more than 500 screen-browsable clip-art images. $495 retail. Also available for DOS and OS/2. Lotus Development Corp., 55 Cambridge Parkway, Cambridge, MA 02142. 617/577-8500.

Visual Horizons provides ready-made slides and overheads or customizes stock presentation graphics. The company also outputs slides from users' computer files for $9 each. Powerful presentation tools for communication excellence for over 20 years. Call for free 56-page color catalog. Visual Horizons, 180 Metro Park, Rochester, NY 14623. 716/424-5300 or Fax 716/424-5313.

Professional Organizers

If you're having trouble getting started, consider professional help to get organized.

National Association of Professional Organizers provides the group's membership directory for $15. 1163 Shermer Rd., Northbrook, IL 60062. 708/272-0135.

Project Management Software (See also Personal Information Manager (PIM)

Keep track of the many details of a project while gaining a big-picture perspective with project management software.

AEC Information Manager is a project-oriented database for the Macintosh designed to organize project information from start to finish ($695). *FastTrack Schedule* generates presentation-quality Gantt charts ($235). *FastTrack Resource* allocates people, equipment, and facilities to projects ($249). AEC Software, 22611 Markey Ct., Bldg. 113, Sterling, VA 22170. 800/346-9413 or 703/450-1980.

CA-SuperProject for Windows allows multiple views of the same project or different projects simultaneously. Includes comprehensive resource management, multiple project linking, extensive analysis, graphing, reporting, and more. $895 for Windows. Available for DOS and VAX. Computer Associates Int'l, Inc., 1 Computer Associates Plaza, Islandia, NY 11788. 800/225-5224 or 516/227-3300.

Texim Project provides multi-projecting including detailed scheduling options, cost accounting, statistics for risk management, and extensive resource management functions. Graphics and color capabilities. Handles unlimited numbers of projects, activities, resources, and calendars. Single PC version, $1,295; LAN version, $2,495. Texim, Inc., 833 Portland Ave., St. Paul, MN 55104. 612/290-9627.

Recycled Products, Environmental Issues, and Miscellaneous Supplies

We don't want to suffer environmental overload in our zeal to SURVIVE Information Overload.

Ad-A-Tab card tabs affix to business cards so they can be added to a rotary file. DO-IT Corporation, PO Box 592, South Haven, MI 49090. 800/426-4822 or 616/637-1121.

Buyers Market allows U.S. consumers to pick and choose 90 categories of direct mail to screen out by giving a thumbs up or down. The resultant "suppression list" is provided to companies free of charge. Lists of consumers giving thumbs up to a category are rented to companies in that category. Equifax, Inc., 801 Penn Ave. NW, Suite 350, Washington, DC 20004. 800/289-7658 or 202/638-0277.

DMA Mail Preference Service offers a free service to remove an individual's name and address from direct-mail lists. More than 1 million people have signed up for the service. Send your request in writing to: Direct Marketing Association, 11 West 42nd Street, PO Box 3861, New York, NY 10163-3861. 212/768-7277.

EarthWise is a line of office products with paper made from recycled fibers. Includes file folders, office paper, storage boxes, 3-ring binders, record books, and more. Available in office supply stores. Esselte Pendaflex Corp., 71 Clinton Rd., Garden City, NY 11530. 516/873-3239.

Quill offers a wide array of recycled office products in its mail-order catalog. Includes bags, data binders, envelopes, file folders, labels, office paper, pads, trays, and more. Quill Corp., PO Box 4700, Lincolnshire, IL 60197. 708/634-4800. Fax 708/634-5708.

Telephone Services

Maximize the benefits of this easy-to-use technology to disseminate rapidly changing information.

ABCvoice is a PC-based voice mail, auto-attendant, and information retrieval system. Allows callers to use touch-tones in response to voice prompts to request and receive information automatically, 24 hours a day. Handles multiple lines simultaneously with up to 6 lines in one computer. Network compatible. Versicom Corp., 316 Regency Ridge, Dayton, OH 45459. 513/438-3700.

Allnet Call Delivery service records your voice message and delivers it to virtually any phone. Allnet Communication Services, Inc., 30300 Telegraph Rd., Birmingham, MI 48010. 313/647-6920.

AT&T Message Service enables you to record a brief message in your own voice and have it delivered to virtually any phone in more than 150 countries at a delivery time you specify. You may request a response at no extra charge. The service is billed to your credit card. To access the service, call 800/562-6275 from the United States or connect via a USA Direct operator from outside the United States.

AT&T TeleConference Service includes a variety of options for teleconferences with multiple callers. No special equipment required. Conference host schedules TeleConference call by dialing an 800 number to set up the conference. Some options allow 500+ conferees. Standard tariffed charges apply. AT&T TeleConference Service, 55 Corporate Dr., Room 14B24, Bridgewater, NJ 08807. 800/232-1234.

Logos is a state-of-the-art call forwarding system as easy to use as an answering machine. Automatically forwards calls to anywhere in the world. Allows you to create a personal voice mail system using your answering machine. $195. Logotronix Inc., 1877 Broadway, Suite 405, Boulder, CO 80302. 800/442-4887 or 303/443-3975.

SkyTel services alert you instantly to your messages nationwide and internationally. *SkyPager* receives numeric messages; *SkyWord* allows you to get concise written messages; *SkyTalk* receives voice mail messages. Also available is *Message Card*, a credit card style message receiver. Sky-Tel Corp, 1350 I Street NW, Suite 1100, Washington, DC 20005. 800/456-3333 or 202/408-7444.

Time Planners and Personal Organizers

Convert these binders into a Master Control System, then combine it with high-technology solutions.

Day Runner organizers are known for their elegance, functionality, and durability and are available in a variety of sizes and styles in office products stores and other retailers. Special pages track projects, people, finances, and more. *Time Plus* software and train-the-trainer available. Over 5 million users worldwide. From $45 to $250. Day Runner, Inc., 2750 Moore Ave., Fullerton, CA 92633. 714-680-3500.

Day-Timers has a variety of notebook and binder-based calendar systems in an assortment of sizes with special-purpose pages to track projects, phone calls, meetings, and more. Over 4 million users worldwide. Sold exclusively by direct mail. Day-Timers, Inc., One Day-Timer Plaza, Allentown, PA 18195. 215/395-5884.

Franklin Day Planner offers a variety of time planners designed by experts to help you balance your personal values with your goals and priorities. Also offers custom binders, accessories, and other products with company name/logo. Franklin International Institute, Inc., PO Box 25127, Salt Lake City, UT 84125. 800/654-1776 or 801/975-1776.

Time/Design offers the binder-based *Management System,* which has won design awards in Europe. The heart of the system is the DataBank, with 10 sections for major categories. Also offers *TimeBase* and *NameBank* software—see Contact Management and Electronic Calendar. Time/Design, 265 Main St., Agawam, MA 01001. 800/637-9942 or 413/789-9633.

Sharp Wizard electronic organizers come in a variety of models featuring typewriter style keyboard, 40-character by 8-line screen, built-in outliner, calendar, scheduler, business card file, phone directories, and built-in help screens. Data can be transferred to and from desktop computers. Under $400. Sharp Electronics Corporation, Sharp Plaza, Mahway, NJ 07430. 800/321-8877.

Index

A

ABCfax, 242
ABCvoice, 252
Abstract images, 176
ACT program, 238
Ad-A-Tab, 251
Address Book Plus, 236–37
Adler, Ralph, 167–68
AEC Information Manager, 248
Agenda software, 120, 247
Alesandrini and Associates, 233
allCLEAR, 246
Allnet Call Delivery, 141, 250
Almanacs, 112
American Society of Training and
    Development, 171, 203
AM Newsbreak, 188
Amoco Production Company, 96
Amos, Wally "Famous," 32, 41
Analogical images, 176
Analogizing, 182
Annenberg School of
    Communications, 134
Apple Computer, 6, 143
Arnum, Eric, 67
Arthur D. Little Online, 116
Asimov, Isaac, 186
Aspects software, 145, 244
Association directory, 111–12
ASTD; *see* American Society of Training
    and Development
AT&T, 97
AT&T EDI, 240
AT&T Enhanced FAX, 242
AT&T Mail, 241
AT&T Message Service, 141, 252
AT&T TeleConference Service, 252
AT&T Toll-Free 800 Directory, 115
ATOSIG program, 119

Audiocassette news service, 190
Audio-visual sources, 233
Autographix Overnight Slide Service,
    249
Automatic voice-fax system, 5
Automation, telephone process,
    142–43

B

Babcock, Miles, 137–38, 191
Balfour, Arthur, 201
Bar-coded micrographics, 70
Battram, Beth Lewis, 5
Bibliographic databases, 116
Big picture, 91–93
    action plan, 111–25
    learning, 180–81
    thinking process, 109–11
    writing, 205–6
Big-picture map, 8
Bolles, Richard, 41–42
Booth, Mac, 95
Borland Company, 5, 62
Bottom-Line Business Writing (Fielden
    and Dulek), 203
Brain research, 109–10
Briefings, 3
Broadband thinking, 121–25
BRS Information Technologies, 247
Bulletin board services, 117
Burns, Anthony, 43
Burrelle's, 237
    Information Search Service, 120, 244
Burrus, Daniel, 9, 19, 22, 72
Business cards, 77–78
Business Contacts and Information
    Manager, 238

Business directories, 115
Business International, 247
Business Meeting Checklist, 137
Butterfield, Leslie, 103
Buyers Market, 251

C

Calano, Jimmy, 191
Calendar, 26
Calendar software, 234
Calloway, Wayne, 136
Canion, Ron, 42
Canofile 250, 246
Canon CAR System, 246
Capture Lab, 8, 147
CareerTrack, 196, 235
Carnevale, Anthony, 169, 203
Car phones, 142
Casio Computer, 122
CA-SuperProject, 250–51
Categorization of activities, 19–20
CA-UptoDate, 236
CD-ROM resources, 234–35
Cellular telephones, 142
Chapman, James, 136
Cheifet, Stewart, 149
Chunking process, 172–74
Churchill, Winston, 210
Clipping services, 2–3, 187–89,
    235–36
Cognitive learning skills, 171–72
Cognitive thinking, 110
CoLab project, 147
Collaboration, 95–100
    with customers and suppliers, 97–98
    electronic meeting rooms, 147–48
    groupware, 144–47
    in learning, 184–85
    in meetings, 137–50
    by phone, 140–43
Collaborative computing, 8
Collaborative work, 6–8; *see also*
    Teamwork

Color-coding
    files, 74–76
    reading material, 195
COMLINE on Silver Platter, 236
Communication Briefings, 239
Communication skills, 169
Companies
    big-picture analysis, 48–55
    cross-functional map, 50–51
    employee training, 159–85
    organizational chart, 49–50
    priorities, 37–38
    process map, 51–52
    vision, 39–41
Company Intelligence, 118, 247
Company profiler, 112–13
Complete Speaker's Almanac, 115
CompuServe, 66, 117, 119, 149, 190,
    248
CompuServe Almanac, 118
Computer Assisted Retrieval system,
    70
Computerized project management,
    103–9
Computer printouts, 67
Computer virus, 120
Confucius, 181
Consumer Electronics Forum, 149
Contact management resources,
    236–37
Corporate Visions, 235
Covey, Stephen, 44, 47
Crandall, Robert, 79, 136
Cross-functional map, 50–51
Current Awareness Service, 190
Customers, collaboration with, 97–98
Customer service
    company-wide scope, 94–95
    electronic mail use, 65–66
    fax machine use, 62

D

D&B Business Information Report, 247
D&B Express, 245

Dartmouth Research, 2, 6
Day Runner, 253
Day-Timers, 253
Decision making, 133
Dialog Information Services, 117, 119, 190, 248
Dictionaries, 114
Digests, 3
Direct Marketing Association Mail Preference Service, 86
Directory of Associations, 112
Disclosure database, 118
Disney, Walt, 216
DMA Mail Preference Service, 251
Drake, Neil, 145
Dulek, Ronald, 203
DuPont Corporation, 63

E

EarthWise, 251
Eastman Chemical Company, 96
EDI; *see* Electronic data interchange
Effectiveness, 43–44
Efficiency, 43–44
Einstein, Albert, 183
Electronic customer forums, 98
Electronic databases, 115–21
Electronic data interchange, 67–68
resources, 238–39
Electronic Data Systems, 8, 147
Electronic forum, 5–6
Electronic libraries, 4
Electronic mail, 62–66
resources, 239
Electronic Mall, 245
Electronic master control system, 22–24
Electronic meeting rooms, 147–48
Electronic meeting systems and services, 240
Electronic paper, 10, 60–61
Employee retention, 163
Employee training
impact assessment, 168

Employee training—*Cont.*
improving, 164–69
instructional technology, 176–78
investing in, 161–64
scope, 159–60
teaching how to learn, 169–72
white collar workers, 169
End-result thinking, 44
Environmental services, 249–50
Equifax Inc., 86
Esselte Pendaflex, 73, 74
European Corporations CD, 236
Executive Book Summaries, 195, 239
Executive News Service, 190
Executive summaries, 195
expEDIte, 240

F

F3 Forms Automation System, 243
Face-to-face communication, 6
Facsimile services, 240–41
Fax machines, 61–62
FaxPump, 62, 242
Federal Express, 17–78
Fiber optics metaphor, 91–93
Fielden, John, 203
Fifth Discipline (Senge), 184
Filing systems, 70–76
Filippo, Lou, 78
Findex, 119
Find It Fast, 114
First Release, 190
FISCAL Directory of Fee-Based Information Services in Libraries, 121, 245
Fitzgerald, Jim, 62
Focused thinking, 110
Ford Motor Company, 167
Forms software, 241
Forum Corporation, 40
Forums
customer, 98
electronic, 148–50
Foster, Richard, 215

Franklin Day Planner, 34, 253
Freelance Graphics, 250
Free (or nearly free) products and
    services, 241–42
Fuller, R. Buckminster, 32, 122
Functional integration, 161–62

G

Gateway services, 17–18
General BusinessFile, 237
General Electric, 97
General Motors Saturn plant, 178
Gilder, George, 1, 91
Gill, Daniel, 79, 142
Glass, David, 38
Global Village Communications, 98
Glossbrenner, Alfred, 116
Goethe, W., 182
Gordon, Jack, 6
Grace, J. Peter, 79, 136
Graham, Diane, 164
Graphics
    media, 134–37
    in meetings, 132–37
Group consensus, 133
GroupSystems, 241
Groupware, 144–47, 242
Grove, Andrew, 42, 79, 136
Guidebooks, 114

H

Handbooks, 114–15
Hand-held time-planner, 20–24
Harley-Davidson, 39
Harris, Louis, 40
Harrison Conference Services, 150
Harvard Business Review, 4, 33, 38, 116,
    158
Hay Group, 39
Herman, Roger, 188
Hewlett-Packard, 6, 99–100
Hey, Kenneth, 123
Higgins, A. Foster, 39
Holmes, Oliver Wendell, 91

Hoover's Handbooks, 112–14, 239
Hotelcopy Faxmsil, 243
Hovdey, Suzanne, 161
How to File, 243
How to File and Find It, 73, 244
How to File Guide, 73
Hughes Aircraft, 65
Hurley, John, 160
Hypermedia, 197
Hypertext, 197
HyperView Systems, 177

I

IBM, 6
    PROFS system, 63
IBM *Current*, 249
Illustrated dictionaries, 114
Image processing, 68–70
Images, 106–7
    automatic quality, 108
    energizing impact, 109
    global nature, 108–9
    types, 175–76
Industry Week, 42
Inferential Focus Briefing, 245
Information
    big picture thinking, 109–11
    big picture view, 91–93
    collaborative work, 6–8
    deceptive, 16–18
    from employees and customers, 4–6
    handling paperwork, 9–10
    increasing amount of, 93
    macro view, 111–15
    networking, 88
    online resources, 115–21
    from reading, 186–202
    reference books, 111–15
    relevant, 2–4
    representations of, 183
    right medium, 16
    scheduling issue, 8–9
    sharing, 93–94
    teamwork and collaboration, 95–100
    visualization, 105–9

Information and research services,
242–43
Information Anxiety (Wurman), 12
Information guidebooks, 114
Information overload
amount, 1–2
and electronic mail, 63–65
escaping, 10–12, 30–31
failure to recognize, 15–16
fiber optics metaphor, 91–94
from meetings, 128–57
review of, 14–15
self-test, 28–30
signs of, 12–14
Information Please Almanac, 112
Information Search Service, 120
Information sifter, 120
Information Visualizer Project, 4
Informed, 243
Ingersoll-Rand, 163
INSPIRATION, 248
Instant Update, 244
Instructional designers, 165–66
Instructional Science, 175
Instructional technology, 176–78
Interconn, 240
International Data Corporation, 61
Interpersonal learning skills, 172
Investment, return map, 6

J

Japan Technology, 247
JetForm Design, 243
Johnson, Samuel, 111
Journal of Mental Imagery, 175
Junk mail, 86
Just-in-time performance support, 167

K

Kanter, Rosabeth Moss, 18
Keeping Good People (Herman), 188
Kelley, Robert, 40

King, Reatha, 109
Knowledge Navigator, 143–44, 150
Kurtzig, Sandra, 43, 136

L

Laserdiscs, 234–35
Lawler-Demitros, Kathleen, 39
Learning
action plan, 178–85
cognitive strategies, 180–84
by collaboration, 184–85
employee training, 159–60
instructional technology, 176–78
motivation, 158
objectives, 166
principles, 172–76
psychology of, 170
theory and basic skills, 168–69
and visualization, 174–76
Learning hierarchy, 183–84
Learning skills, 169–72
Learning style, 178–80
Learning Styles Inventory, 171
Legal databases, 118
Leonardo da Vinci, 33
*Lesko's Information Power,*114, 240
Lewis, T. Reid, 145
Library Research Group, 190
Logos, 252
Lotus CD/Corporate, 237
Lotus Development Corporation, 15
Lotus Notes, 244
Luce Press Clippings, 237

M

Mackay, Harvey, 76
Macro reference shelf, 111–15
Magazine Index, 60, 247
Mail order services, 243
Management forms, 104
Managers, communicating vision,
39–40
Mansfield, Mike, 215

Manzi, Jim, 14
Mapping, 6, 8
  big picture analysis, 48–55
  return map, 99–100
Market Force Plus, 238–39
Marshall, Patricia, 138–40
Master control system, 9
  categorizing activities, 19–20
  context of priorities, 34
  electronic, 22–24
  as information channel, 24
  management forms, 104
  maximizing use of, 24–28
  for meetings, 137–40
  need for, 18–19
  time-planner, 20–24
Master Graphics, 135
Mastering the Information Age
  (McCarthy), 196
Math skills, 169
McCarthy, Michael, 196
McDermott, Robert, 10, 69, 161
McDonnell Douglas, 130
MCI Mail, 65, 241
McLaughlin, Edwina, 83–85
MCS; see Master control system
Media Industry Newsletter, 108
Medical database, 118
Meeting Maker, 246
Meeting management, 244
Meetings
  action play, 150–55
  collaborative work, 137–40
  cost of, 130–31
  executive views, 136
  and information overload, 128–
    29
  knowledge navigator, 143–44
  unnecessary, 129–30
  use of graphics, 132–37
  Wharton study, 132–34
Megatrends (Naisbitt), 12
Megatrends 2000 (Naisbitt), 123
Memory, 172–74
Mental outline, 9

Merck Corporation, 164
Messaging systems, 141
Meyers-Briggs Type Indicator, 171
Michigan Information Transfer Source,
  121, 245
Microfiche, 69–70
Microfilm, 69–70
Micrographics and electronic filing,
  244
Miller, George, 173
Miscellaneous supplies, 249–50
Moody's Corporate Profiles, 117
Motorola Corporation, 8, 68, 161, 166,
  168, 181
Mozart, W. A., 44
Multidisciplinary work teams, 96–97

N

Naisbitt, John, 12
NameBank, 239
National Association of Professional
  Organizers, 250
National Directory, 112
National Directory of Address and
  Telephone Numbers, 240
National Zip + 4 Diskette Processing
  Service, 244
Navistar, 97
NCR, 177
Nesting, 183
Networking
  with customers, 98
  online, 148–50
New Republic, 206
News-Express, 188
Newspapers, 192–94
Newstrack Executive Tape Service, 190,
  196, 236
Newswire services, 188
New York Times, 194
Nightingale-Conant, 196, 236
Northrup Corporation, 70

O

Office filing systems, 70–76
Online bulletin board, 5–6
Online databases, 244–45
Online forums, 119
Online information search service, 190
Online map/navigation chart, 119–20
Online networks, 148–50
Online searching, 3–4
Online services and aides, 115–21,
   245–46
OnTime, 236
Optical laserdisc, 69
Organization map, 49–50
Ounjian, Marilyn, 61, 79
Outlining and diagraming software,
   246–47

P

Panza, Carol, 49–50, 51
Paperwork
   action plan, 79–8
   business cards, 77–78
   computer printouts, 67
   cut by electronic mail, 65
   and electronic data interchange,
      67–68
   and electronic paper, 60–61
   excess and costs, 58–59
   executive views, 79
   image processing, 68–70
   and job responsibility, 59–60
   office filing systems, 70–76
   overload, 9–10
   phone message slips, 76–77
   self-test, 88–90
Paperwork Reduction Act, 59
Parrish, W. Scott, 68
Parrish-Keith-Simmons, 67–68
Patents, 96
PC Computing, 23
Pennsylvania State University, 166

Performance-based instruction, 166
Performance Evaluation and Review
   Technique/Critical Path Method,
   104
Performance Research Associates, 94
Performance support, 167
Personal information manager, 23–24,
   247
Petty, Bruce, 170
Phone message slips, 76–77
Phone Power (Walther), 140
*Picture This. . . .Your Function, Your
   Company* (Panza), 49
Pilot testing, 98
Planning calendar, 26
Planning Review, 104
PLATO system, 178
Poppa, Ryal, 5, 136, 153, 162–63
Power Graphics, 135
Predicasts F&S Index plus Test, 237
Presentation graphics products and
   services, 247–48
Priorities
   activity analysis, 46–48
   big picture analysis, 48–55
   company, 37–38
   context, 34
   effectiveness versus efficiency,
      43–44
   executive views, 42–43
   external context, 36–37
   individual, 52–55
   internal context, 37
   need for, 41–42
   ordering, 44–56
Process map, 51–52
Procter & Gamble, 177
Productivity
   collaborative, 6–8
   gains from training, 161
Productivity pyramid, 34, 35
Professional organizers, 248
Project management, visualized, 103–4
Project management software, 248–49
Pro-Search, 119, 248

Q

Quality, 163–64
QuickMail, 241
Quill Corporation, 73, 251

R

Radio TV Reports, 238
Reader's Digest Reverse Dictionary, 114
Reading
  action plan, 197–202
  clipping services, 187–89
  reducing load of, 191–96
  speeding up, 196–97
  streamlining, 194–96
  value and reasons for, 198–99
Reading skills, 169
Recordable compact disks, 69
Reference sources, 111–15
Representational images, 175
Return map, 6, 99–100
Rico, Gabriele, 210
Rosenshine, Allen, 43
RQ3S reading method, 198

S

Sales automation resources, 236–37
Salzman, Jeff, 191
Satellite communications, 150
Scheduler software, 234
Scheduling, 8–9
Science database, 118
Sekora, Michael, 122
Senge, Peter, 184
Service companies, economic decline,
  18
Sharp, Seena, 112
Sharp Wizard, 253
Sheedy, James, 163
Short-term memory, 173
Skimming, 201–2
SkyTel, 252

Smith, Louis, 8–9, 163
Software for collaborative computing,
  144–47
Sony Corporation, 6
Source database, 116
Speaker's and Toastmaster's Handbook,
  114–15
Spero, Joan, 40
SQ3R reading method, 197
Standard & Poor's Register, 117
Staples stores, 246
Stata, A. R., 158
Storage Technology, 162–63
STX12, 240
Subscriptions, 191–92
Sullivan, Barry, 43
Sumberg, Brenda, 8, 166, 168
Summaries, 3
Suppliers, 97–98
*Swim With the Sharks Without Being
  Eaten Alive* (Mackay), 76

T

TAPCIS, 248
TeamFocus, 241
Teamwork, 95–100
TechFax, 62
Technology database, 118
TeleMagic, 239
Telephone messaging services, 250
TelePort, 242
Texim Project, 251
Thinking process, 109–10
  broadband, 121–25
  chunking, 172–74
Thomas Register Online, 117
3-D Information Visualizer, 4
3 M Company, 132, 134
TimeBase, 249
Time/Design, 253
Time management
  activity analysis, 46–48
  by effective executives, 33–34

Time Management—*Cont.*
  myths, 2–10
  need for, 32–33
  need for priorities, 41–44
  overplanning problem, 8–9
  paperwork problem, 58–90
  perspectives, 35–37
  self-test, 56–57
  tools, 34–35
  traditional, 1–2
Time management matrix, 45–56
Time-planner, 20–24
Time planners and personal
    organizers, 250–51
Toffler, Alvin, 169–70
TopDown, 249
Top-down reading, 197–202
Top-down writing, 207–11
Trade guides, 115
Training; *see* Employee training *and*
    Learning
Training magazine, 163, 164
Treleaven, Jim, 138
Turnaround through training, 162–63

U

United Research, 104
United Service Automobile
    Association, 161
Universal Almanac, 112
University of Minnesota, 134
USA Today, 194

V

Vance, Mike, 216
Video conferences, 150
Video-tele-computer conferencing,
    143–44
Virus, computer, 120
Vision
  communicating, 39–40

Vision—*Cont.*
  company, 39
  reasons for, 41
VisionQuest, 241
Visual Horizons, 250
Visualization
  automatic quality, 108
  and concreteness, 174–76
  energizing impact, 109
  global view of projects, 108–9
  by images, 106–7
  importance of, 105–9
  memorable quality, 107–8
  project management, 103–4
  return map, 99–100
  use of graphics, 132–37
Voice-fax computer technology, 148
Voice-fax response, 5
Voice mail, 141
Vu/Clip, 238

W

Wall Street Journal, 128, 194
Wal-Mart Stores, 38
Walther, George, 140
Walton, Sam, 150
Washington Post, 130
Ways of Thinking test, 178–80
Welch, Jack, 136
Wetherbe, James, 93–94
Wharton Business School Study,
    132–34
What Color Is Your Parachute? (Bolles),
    41
Whistler, James, 44
White collar workers, 169
Whitman, Walt, 196
Wiggenhorn, Bill, 161
Wilson, Flip, 153
Wittrock, Merlin, 202
Work teams, 96–97
Writers Market, 114

Writing
  action plan, 207–11
  big-picture, 205–6
  brevity, 204–5
  problems with, 203–4
Writing the Natural Way (Rico), 210
Wurman, Richard, 12

X

Xerox Corporation, 6
Xerox Palo Alto Research Center, 4, 147

Y

Young, John, 43